Fundamentals
of Programmable Logic
Controllers, Sensors,
and Communications

Jon Stenerson

REGENTS/PRENTICE HALL, Englewood Cliffs, New Jersey 07632

Library of Congress Cataloging-in-Publication Data

Stenerson, Jon.
  Fundamentals of programmable logic controllers, sensors,
  and communications / Jon Stenerson
    p.   cm.
    Includes index.
    ISBN 0-13-726860-2
  1. Programmable controllers I. Title.
  TJ223.P76S74   1994
  629.8'9--dc20   92-41671
                                       CIP

Production Editor: *Mary Carnis*
Supplement Acquisitions Editor: *Judy Casillo*
Acquisitions Editor: *Holly Hodder*
Prepress Buyer: *Ilene Levy-Sanford*
Manufacturing Buyer: *Ed O'Dougherty*

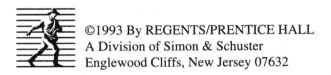 ©1993 By REGENTS/PRENTICE HALL
A Division of Simon & Schuster
Englewood Cliffs, New Jersey 07632

Printed in the United States of America

10 9 8 7 6 5 4 3 2 1

ISBN 0-13-726860-2

Prentice-Hall International (UK) Limited, *London*
Prentice-Hall of Australia Pty, Limited, *Sydney*
Prentice-Hall Canada Inc., *Toronto*
Prentice-Hall Hispanoamericana, S.A., *Mexico*
Prentice-Hall of India Private Limited, *New Deli*
Prentice-Hall of Japan, Inc., *Tokyo*
Simon & Schuster Asia Pte. Ltd., *Singapore*
Editora Prentice-Hall do Brasil, Ltda., *Rio de Janero*

This book is dedicated to Jane and Jeff for their patience and support and also Dick and Virg for enabling me to pursue my interests in my work.

Jon Stenerson

Contents

Preface   xi

Chapter 1 Overview of Programmable Logic Controllers   **1**

Objectives   1
History of PLCs   2
PLC Components   6
Central Processing Unit   7
Memory   8
PLC Programming Devices   12
Power Supply   16
Input Section   17
Output Section   24
PLC Applications   27
Questions   30

Chapter 2 Overview of Number Systems   **33**

Objectives   33
Decimal   34
Binary Numbering System   35
Binary Coded Decimal System   37
Octal   37
Hexadecimal   38
Questions   43

Chapter 3 Fundamentals of Programming   **45**

Objectives   45
Contacts   46
Coils   47
Ladder Diagrams   47
Multiple Contacts   52
Branching   54
Special Contacts   57
PLC Scanning and Scan Time   60
Questions   62

Chapter 4 Timers and Counters 65

Objectives 65
Timers 66
Allen Bradley PLC-2 Timers 68
Gould Modicon Timers 70
Omron Timers 71
Square D Timers 73
Siemens Industrial Automation Timers 74
Cascading Timers 77
Counters 77
Allen Bradley PLC-2 Counters 78
Gould Modicon Counters 80
Omron Counters 80
Square D Counters 82
Siemens Industrial Automation Counters 83
Programming Hints 85
Questions 88
Additional Exercises 92

Chapter 5 Sensors and Their Wiring 93

The Need For Sensors 93
Typical Applications 94
Sensor Types 96
Contact vs. Noncontact 96
Digital vs. Analog 96
Digital Sensors 97
Optical Sensors 97
Electronic Field Sensors (Field Sensors) 100
Capacitive Sensors 104
Sensor Wiring 105
Load-Powered Sensors 105
Line-Powered Sensors 107
Other Transducers 108
Installation Considerations 112
Electrical 112
Mechanical 112
Typical Sensor Applications 113
Questions 117

Chapter 6 Input/Output Modules and Wiring   **119**

I/O Modules   119
Digital (Discrete) Modules   120
Analog Modules   125
Resolution in Analog Modules   126
Remote I/O Modules   127
Communication Modules   128
Position Control Modules   129
Vision Modules   131
Bar-Code Modules   133
PID Modules   133
Fuzzy Logic Modules   135
Radio-Frequency Modules   136
Operator Input/Output Devices   137
Speech Modules   138
Questions   140

Chapter 7 Arithmetic Instructions   **141**

Objectives   141
Introduction   142
Allen Bradley Arithmetic Instructions   143
Gould Modicon Arithmetic Instructions   148
Omron Arithmetic Instructions   150
Square D Arithmetic Instructions   154
Siemens Industrial Automation Arithmetic Instructions   159
Questions   161

Chapter 8 Advanced Programming   **163**

Objectives   163
Sequential Control   164
Allen Bradley Sequencer Instructions   166
Shift Resister Programming   169
Stage Programming   170
Step Programming   174
Fuzzy Logic   177
State Logic   184
Questions   188
Additional Exercises   188

Chapter 9 Overview of Plant Floor Communication   189

Objectives   189
Introduction   190
Levels of Plant Communication   190
Device Level   192
Cell Level   193
Types of Cell Controllers   196
Area Control   199
Local Area Networks (LANs)   199
Topology   200
Cable Types   202
Access Methods   204
Host Level   208
Questions   209

Chapter 10 Installation and Troubleshooting   211

Objectives   211
Installation   212
Enclosures   212
Wiring   213
Grounding   215
Handling Electrical Noise   217
Noise Suppression   217
Noise Isolation   218
PLC Troubleshooting   219
Questions   223

Appendix A Allen Bradley PLC-5 and
**   SLC 500 Instructions   225**

Logical Addressing   225
I/O addressing   226
Examine-On   228
Examine-Off   228
Output Energize   229
Output Latch   229
Output Unlatch   229
Immediate Input   230
Immediate Output   231
Timer On Delay   231

Timer Off Delay   233
Retentive Timer On   234
Counters   235
Master Control Reset Instructions   238
Arithmetic Instructions   239
Limit   241
Compute   243
Square Root   248
Standard Deviation   249
Average Instruction   250
Logical Operators   251
Number System Conversion   255

Appendix B I/O Device Symbols   **257**

Index   **269**

Preface

The programmable logic controller has become an invaluable tool in industry. The use of programmable controllers is helping transform American industry. There is a huge need for trained personnel who can program and integrate programmable logic controllers. The integration of the programmable logic controller is the key.

I began writing this text for my students when I was unable to find a practical, affordable text that used a generic approach.

Each chapter begins with a generic approach to the topic. Each topic is clearly explained through the use of common, easy to understand, generic examples. Examples are then shown for specific controllers. The brands covered are Allen Bradley, Gould Modicon, Omron, Siemens Automation, and Square D Corporation. There are many illustrations and practical examples. Each chapter has objectives and questions. Some chapters have programming exercises. Software is available free for those instructors that adopt this text. Contact REGENTS/Prentice Hall to acquire the software. The software consists of TISOFT ladder logic programming software, and tutorials on stage programming and state logic. This software is courtesy of Siemens Automation Inc. and Adatek. There are additional learning exercises in several chapters that utilize the software. After completing the reading and exercises, the reader will be able to easily learn any new programmable controller. The text also covers new and emerging technologies such as fuzzy logic, radio frequency, as well as new programming techniques such as step, stage and state logic programming. The book is not intended to replace the technical manual for the specific programmable logic controller. The book does however, explain the common instructions so that the reader will then be able to efficiently learn the use of new instructions from any technical manual.

Chapters 1-4 provide the basic foundation for the use of PLCs. Chapter 1 focuses on the history and fundamentals of the PLC. Chapter 2 covers number systems. Chapter 3 covers contacts, coils, and the fundamentals of programming. Chapter 4 focuses on timers, counters, and logical program development.

Chapter 5 covers industrial sensors and their wiring. The chapter focuses on types and uses of sensors. Sensors covered include optical, inductive, capacitive, ultrasonic, and thermocouples. The wiring and practical application of sensors is stressed.

Chapter 6 covers I/O modules and wiring. The chapter covers digital and analog modules, communication modules, position control modules, barcode modules, radio frequency modules, fuzzy logic modules, speech modules, and others.

Chapter 7 covers arithmetic instructions. The common arithmetic instructions including add, subtract, multiply, divide, and compare are covered. In addition logical operators, average, standard deviation, and number system conversion instructions are covered.

Chapter 8 covers advanced programming. Sequential logic, shift registers, step logic, stage logic, fuzzy logic, and state logic are all covered. Several of these programming techniques are very new, but will rapidly become prevalent.

Chapter 9 covers plant communication. Device communications are sure to increase in importance as companies integrate their enterprises. This chapter provides a foundation for the integration of plant floor devices.

Chapter 10 focuses on the installation and troubleshooting of PLC systems. The chapter begins with a discussion of cabinets, wiring, grounding, and noise. The chapter then covers troubleshooting. This chapter will provide the fundamental groundwork for proper installation and troubleshooting of integrated systems.

Appendix A is provided for those readers with Allen Bradley PLC-5s or SLC 500s. The common PLC instructions are explained.

Acknowledgments

I would like to acknowledge the support of Adatek, Allen Bradley, Giddings & Lewis Electronics, Omron, Square D, and Siemens Automation Inc. Without their support much of what we have accomplished at our college would not have been possible.

I would also like to thank the following reviewers for their efforts:

Mr. Terry Fleischman, Fox Valley Technical College, Appleton Wi
Mr. Frank Gergelyi, New Jersey Institute of Technology, Newark, NJ
Mr. Nazar Karzay, Indiana Vocational Technical College, Evansville , IN

Chapter 1

Overview of
Programmable Logic Controllers

In a very short period of time, programmable logic controllers (PLCs) have become an integral and invaluable tool in industry. In this chapter we will examine how and why PLCs have gained such wide application. We will also take an overall look at what a PLC is.

OBJECTIVES

Upon completion of this chapter, you will be able to:

Explain some of the reasons why PLCs are replacing hardwired logic in industrial automation.

Explain such terms as <u>ladder logic</u>, <u>CPU</u>, <u>programmer</u>, <u>input devices</u>, and <u>output devices</u>.

Explain some of the features of a PLC that make it an easy tool for an electrician to use.

Explain how the PLC is protected from outside disturbances.

Draw a block diagram of a PLC.

Explain the types of programming devices available.

History of PLCs

The programmable logic controller may be the best example ever, of taking an existing technology and applying it to meet a need. In the 1960s and 1970s, industry was beginning to see the need for automation. Industry saw the need to improve quality and increase productivity. Flexibility had also become a major concern. Industry needed to be able to change processes quickly to meet the needs of the consumer.

The Old Way

Imagine an automated manufacturing line in the 1960's and 1970's. There was always a huge wiring panel to control the system. The wiring panel could cover an entire wall. Inside the panel were masses of electromechanical relays. These relays were all hardwired together to make the system work. Hardwiring means that an electrician had to install wires between the connections of the relays. An engineer would design the logic of the system and the electricians would be given a blueprint of the logic and would have to wire the components together. There were hundreds of electromechanical relays in a system before programmable logic controllers were developed.

The drawing that the electrician was given was called a ladder diagram. They were called ladders because they resemble a ladder in appearance. The ladder showed all the switches, sensors, motors, valves, relays, etc., that would be in the system. It was the electrician's job to wire them all together (see Figure 1-1).

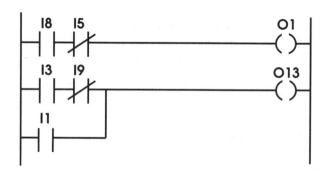

Figure 1-1. PLC ladder diagram. Note the similarity in appearance to a ladder. There are three rungs in this example. Power is represented by the left and right uprights. There are five inputs and two outputs in this example. Inputs are on the left, outputs at the right of each rung. Appendix B shows examples of electrical symbols.

It is not hard to imagine that the engineer made a few small errors in his/her design. It is also conceivable that the electrician may have made a few errors in wiring the system. It is also not hard to imagine a few bad components. The only way to see if everything was correct was to run the system. Systems are normally not perfect on the first attempt. Troubleshooting was done by running the actual system. This was a very time-consuming process. You must also remember that no product could be manufactured while the wiring was being changed. The system had to be disabled for wiring changes. This means that all of the production personnel associated with that production line were without work until the system was repaired.

After the electrician had completed the troubleshooting and repair, the system was ready for production.

Disadvantages of the Old Way

One of the problems with this type of control is that it is based on mechanical relays. Mechanical devices are usually the weakest link in systems. Mechanical devices have moving parts that can wear out. If one relay failed, the electrician might have to troubleshoot the whole system again. The system was down again until the problem was found and corrected.

Another problem with hardwired logic is that if a change must be made, the system must be shut down and the panel rewired. If a company decided to change the sequence of operations (even a minor change), it was a major expense and loss of production time while the system was not producing parts.

The First Programmable Controllers

General Motors saw the need for a replacement for hardwired control panels. Increased competition forced the automakers to improve manufacturing performance in both quality and productivity. Flexibility, rapid changeover, and reduced downtime became very important.

GM realized that a computer could be used for logic instead of hardwired relays. The computer could take the place of the huge, costly, inflexible, hardwired control panels. If changes in the system logic or sequence of operations were needed, the program in the computer could be changed instead of rewiring. Imagine eliminating all the downtime associated with wiring changes. Imagine being able to completely change how a system operated by simply changing the software in a computer.

The problem was how to get an electrician to accept and use a computer. Systems are often very complex and require complex programming. It was out of the question to ask plant electricians to learn and use a computer language in addition to their other duties.

GM's Hydromatic Division realized the need and acted. In 1968, Hydromatic wrote the design criteria for the first programmable logic controller. [Note: There were companies already selling devices that performed industrial control, but they were simple sequencing controllers—not PLCs as we think of them today (see Figures 1-2 and 1-3).]

The specifications required that the new device be a solid-state device (electronic, not mechanically operated). It had to have the flexibility of a computer. It had to be able to function in an industrial environment (vibration, heat, dirt, etc.) with the ability to be reprogrammed and used for other tasks.

The most important design criterion was that it had to be easily programmed and maintained by plant electricians and technicians. GM solicited interested companies and encouraged them to develop a device to meet the design specifications.

Gould Modicon developed the first device that met the specifications. The key to the programmable logic controllers' success is that the device is programmed using the same language that electricians already used: ladder logic. Electricians and technicians could very easily understand these new devices because the logic looked similar to the old logic they had always worked with. They did not have to learn a new programming language. In fact, the ladder looks almost like the electrical diagrams they used to troubleshoot the systems.

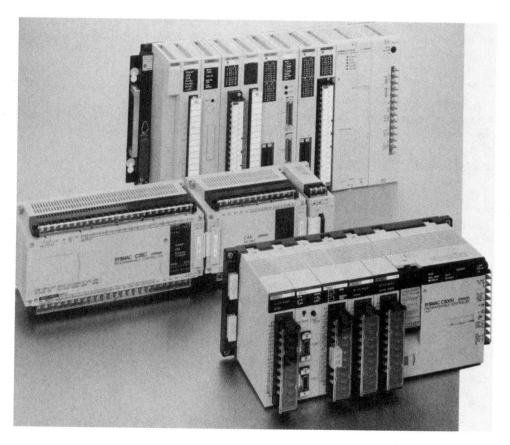

Figure 1-2. This figure shows three different models of Omron PLCs. Most manufacturers offer a wide variety of PLC models from small to large. Courtesy of Omron Electronics.

PLCs made troubleshooting a very easy task also. All of the inputs and outputs could be viewed on the screen of the programming unit. The problem in the system could be found very quickly (see Figure 1-4). If the process needed to be changed, the new ladder could be written off-line. The system could be running while the electrician was making changes in the ladder diagram. The system would only have to be shut down for a few minutes while the new ladder (logic) was downloaded to the PLC (no rewiring).

The ease of change is very important in manufacturing. This capability allows for rapid production or model change. This even makes it possible that each product through a system could be custom made to meet an individual consumer's need. Just-in-time and other new manufacturing techniques require smaller lot sizes of production, which mean more changeovers in production. Speed and flexibility are crucial in this environment. PLCs are the perfect control device for this need. The program in the PLC (ladder diagram) controls how the PLC runs the manufacturing process. If the program is changed, the PLC can completely change the process.

Figure 1-3. This figure shows an Allen Bradley SLC 500 PLC and modules. Courtesy of Allen Bradley Company, Inc., a unit of Rockwell International.

PLCs were originally called PCs (programmable controllers). This caused some confusion when personal computers became prevalent. To avoid confusion PCs became personal computers and programmable controllers became PLCs (programmable logic controllers). The original PLCs were simple devices. They were simply on/off control devices. They could take input from devices such as switches, digital sensors, etc., and turn output devices on or off. They were very appropriate for simple relay replacement applications. They were not well suited for complex control such as temperatures, positions, pressures, etc. Since the early days, manufacturers of PLCs have added numerous features and enhancements. PLCs have also been given the capability to handle extremely complex tasks such as position control, process control, and other difficult applications. Their speed of operation has drastically improved. Their ease of programming has also improved. Special purpose modules have been developed for such applications as radio frequency communications, vision inspection, and even speech. It would be difficult to imagine a task that a PLC could not handle.

Figure 1-4. This shows the use of a computer being used to troubleshoot a system. Courtesy of Square D.

PLC Components

The PLC is really an industrial computer in which the hardware and software have been specifically adapted to the industrial environment and the electrical technician. Figure 1-5 shows the functional components of a typical PLC. Note the similarity to a computer.

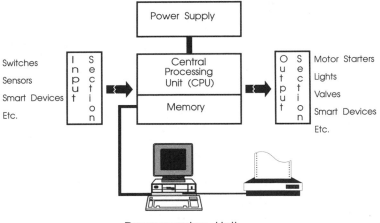

*Figure 1-5. Block diagram of the typical components that make up a
PLC. Note particularly the input section, the output section, and the central
processing unit CPU. The broken arrows from the input section to the CPU,
and from the CPU to the output section represent protection that is necessary
to isolate the CPU from the real-world inputs and outputs. The programming
unit is used to write the control program (ladder logic) for the CPU. It is also
used for documentation of the programmers logic and troubleshooting the
system.*

Central Processing Unit

The central processing unit (CPU) is the brain of the PLC (see Figure 1-6). It contains one or
more microprocessors to control the PLC. The CPU also handles the communication and
interaction with the other components of the system. The CPU contains the same type of
microprocessor that one could find in a microcomputer. The difference is that the program that
is used with the microprocessor is just written to accommodate ladder logic instead of other
programming languages. The CPU executes the operating system, manages memory, monitors
inputs, evaluates the user logic (ladder diagram), and turns on the appropriate outputs.

The factory floor is a very noisy environment. The big problem is the electrical noise that
motors, motor starters, wiring, welding machines, and even fluorescent lights create. PLCs are
hardened to be noise immune.

PLCs also have elaborate memory-checking routines to be sure that the PLC memory has not
been corrupted by noise or other problems. Memory checking is undertaken for safety reasons. It
helps assure that the PLC will not execute if memory is corrupted for some reason. Most
computers do not offer noise hardening or memory checking. There are a few industrial computers that do.

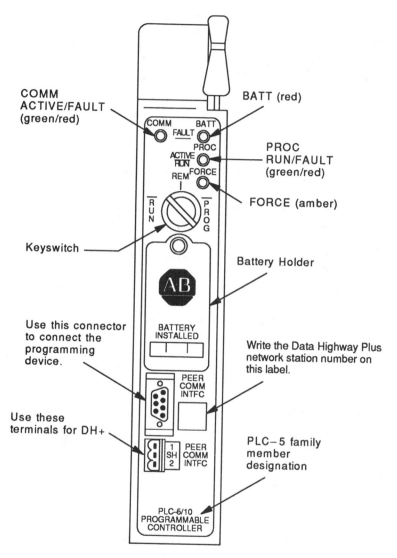

COMM
ACTIVE/FAULT
(green/red)

BATT (red)

PROC
RUN/FAULT
(green/red)

FORCE (amber)

Keyswitch

Battery Holder

Use this connector
to connect the
programming
device.

Write the Data Highway Plus
network station number on
this label.

Use these
terminals for DH+

PLC-5 family
member
designation

Figure 1-6. PLC-5/15 processor module (CPU). Note the many indicators for error checking. Also note the keyswitch for switching modes between run and program mode. Courtesy of Allen Bradley Company, Inc., a unit of Rockwell International.

Memory

PLC memory can be of various types. Some of the PLC memory is used to hold system memory and some is used to hold user memory.

Operating System Memory

Read-Only Memory

Read-only memory (ROM) memory is used by the PLC for the operating system. The operating system is burned into ROM memory by the PLC manufacturer. The operating system controls functions such as the system software that the user uses to program the PLC. The ladder logic that the programmer creates is a high level language. A high level language is a computer language that makes it easy for people to program. The system software must convert the electrician's ladder diagram (high-level language program) to instructions that the microprocessor can understand. ROM is not changed by the user. ROM is nonvolatile memory, which means that even if the electricity is shut off, the data in memory is retained.

User Memory

The memory of a PLC is broken into blocks that have specific functions. Some sections of memory are used to store the status of inputs and outputs. (Note: input/output is typically represented as I/O.) These are normally called I/O image tables. The states of inputs and outputs are kept in I/O image tables. The real-world state of an input is stored as either a "1" or a "0" in a particular bit of memory. Each input or output has one corresponding bit in memory (see Figures 1-7 and 1-8).

Other portions of the memory are used to store the contents of variables that are used in a user program. For example, a timer or counter value would be stored in this portion of memory. Memory is also reserved for processor work areas.

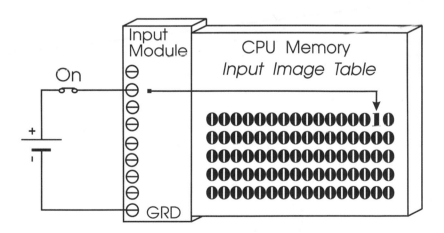

Figure 1-7. How the status of a real-world input becomes a 1 or a 0 in a word of memory. Each bit in the input image table represents the status of one real-world input.

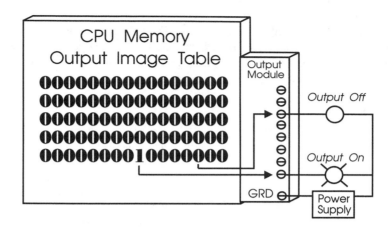

Figure 1-8. How a bit in memory controls one output. If the bit is a 1, the output will be on. If the bit is a 0, the output will be off. (This is active-high logic.)

Memory Maps

Diagrams that show the layout, uses, and location (addresses) of memory are called memory maps. Figure 1-9 shows an example of an Allen Bradley memory map for a PLC-2.

Random Access Memory

Random access memory (RAM) is designed so that the user can read or write to the memory. Ram is commonly used for user memory. The user's program, timer/counter values, input/output status, etc., are stored in RAM.

RAM is volatile which means that if the electricity is shut off, the data in memory is lost. This problem is solved by the use of a lithium battery. The battery takes over when the PLC is shut off. Most PLCs use CMOS-RAM technology for user memory. CMOS-RAM chips have very low current draw and can maintain memory with a lithium battery for an extended period of time, two to five years in many cases.

Nonvolatile Memory

Erasable programmable read-only memory (EPROM) is special ROM. Normally, ROM cannot be altered; it is for reading only. EPROM is nonvolatile memory that can be changed by a user (see Figure 1-10).

The EPROM can be erased and new information (user ladder logic) can be written into it. The EPROM is erased by exposing it to ultraviolet light. The EPROM has a window that is covered. The cover can be removed and the window exposed to an ultraviolet source. This exposure erases the memory. The EPROM can then be reprogrammed. The EPROM is a nonvolatile memory device and does not require battery backup.

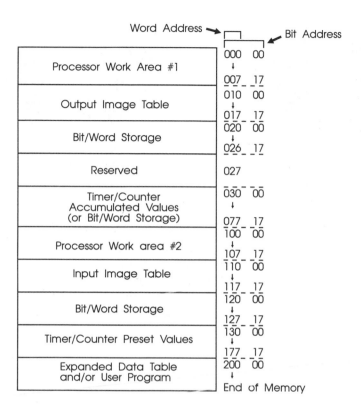

Figure 1- 9. *This figure shows the memory map for an Allen Bradley PLC -2. Note this is the factory configuration. Note that the addresses are in octal.*

Processor work area is used by the processor for its internal operations. These addresses are not available to the programmer for addressing. Note that there are two work areas in this memory map.

Timer/counter accumulated value storage is used to store the present accumulated values for timers and counters.

Timer/counter preset value storage is used to store the user preset values for timers and counters. If the user wants a counter to count up to five this is the area where the value five is stored.

The input /output image table is an area of memory dedicated to I/O data. During every scan each input controls one bit in the input image table. Individual outputs are also controlled by one bit in the output image table.

Figure 1-10. Two kinds of memory: RAM and EPROM. Note the window on the EPROM memory chip. If the window is exposed to ultraviolet light, the memory is erased.

Electrically Erasable Programmable Read-Only Memory

Electrically erasable programmable read-only memory (EEPROM) can function in almost the same manner as RAM memory. The EEPROM can be erased electrically instead of by ultraviolet light. The EEPROM is also nonvolatile memory, so it does not require battery backup. EEPROM modules are available for many PLCs today. They are often small cartridges that can store several thousand bytes of memory (see Figure 1-11). They can be used to store user programs. It is copied to user RAM for execution of the ladder.

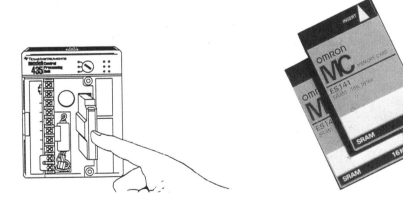

Figure 1-11. Left: shows a memory cartridge for a Siemens Industrial Automation, Inc. PLC. Right: a SRAM (static random access memory) card for a Omron PLC. Courtesy of Omron Electronics and Siemens Industrial Automation, Inc..

PLC Programming Devices

There are many devices that are used to program PLCs. These devices do not need to be attached to the PLC once the ladder is written. The devices are just used to write the user program for the PLC. They may also be used to troubleshoot the PLC.

Dumb Terminal

A dumb terminal is a device that has a keyboard and a monitor. There is no intelligence in the device. This requires that all of the brains for operation of the PLC and for the programming system be located in the PLC. The dumb terminal is just used to send information to the PLC and display the information that is returned from the PLC. A dumb terminal sends out the ASCII equivalent of whichever key was pressed. The output from this device is sent by serial communication to the CPU. The CPU can also send ASCII information back to the dumb terminal. The main advantage of the device is that it is inexpensive and can be used with a variety of devices.

Dumb terminals are not widely used anymore because they cannot upload/download or store programs. Users have also demanded more in the way of documentation and troubleshooting, so the dumb terminal seems doomed for most applications.

Figure 1-12. Allen Bradley programming terminal.

Dedicated Industrial Terminals

These are terminals that have built-in intelligence. They are dedicated to one brand of PLC. In fact, many are dedicated to only a few models of one brand. In many cases, they must be attached to the PLC to be able to program (on-line programming). Some dedicated terminals allow off-line programming. "Off-line" means that the program can be written without being connected to the PLC. The ladder can then be downloaded to the PLC. Dedicated devices can be used to troubleshoot ladder logic while the PLC is running. They can force inputs and outputs on or off for troubleshooting.

The big disadvantage of these terminals is that one is required for each different brand of PLC an industry might have. They are also quite expensive and are only used to program one brand of PLC. For these reasons they are becoming less prevalent.

Hand-Held Programmers

Hand-held programmers are often used to program small PLCs (see Figure 1-13). They are inexpensive and easy to use. Handheld programmers are basically dumb devices also. They must be attached to a PLC to be used. They are handy for troubleshooting. They can easily be carried out to the manufacturing system and plugged into the PLC. Once plugged in they can be used to monitor the status of inputs, outputs, variables, counters, timers, etc. This eliminates the need to carry a large programming device out onto the factory floor. Handheld programmers can also be used to force inputs and outputs on or off for troubleshooting. Handheld programmers are designed for the factory floor. They typically have membrane keypads that are immune to the contaminants in the factory environment.

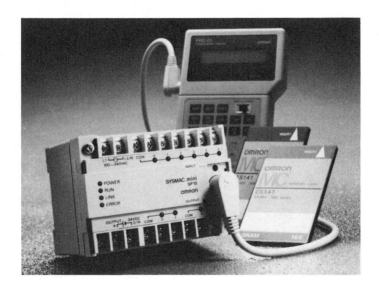

Figure 1-13. This figure shows a hand-held programmer. Note the SRAM memory card. Courtesy of Omron Electronics.

One disadvantage is that these programmers cannot show very much of a ladder on the screen at one time. Some hand-held programmers can also use memory cards. The memory cards can hold programs. The programs can then be downloaded to the PLC.

Chapter 1: Overview of Programmable Logic Controllers

Microcomputers

The microcomputer is rapidly becoming the most commonly used programming device. The same microcomputer can program any brand of PLC that has software available for it. This means that a microcomputer can be used to program virtually any PLC. The microcomputer can also be used for off-line programming and storage of programs. One disk can hold many ladders. The microcomputer can also upload and download programs to a PLC. The microcomputer can also force inputs and outputs on and off.

This upload/download capability is vital for industry. Occasionally, PLC programs are modified on the factory floor to get a system up and running for a short period of time. It is vital that once the system has been repaired the correct program is reloaded into the PLC. It is also useful to verify from time to time that the program in the PLC has not been modified. This can help to avoid dangerous situations on the factory floor. Some automobile manufacturers have set up communications networks that regularly verify the programs in PLCs to assure that they are correct.

The microcomputer can also be used to document the PLC program. Notes for technicians can be added and the ladder can be output to a printer for hardcopy so that the technicians can study the ladder diagram.

Siemens Industrial Automation, Inc. TISOFT is an example of microcomputer software for programming PLCs (see Figure 1-14). TISOFT is a powerful software package that allows off-line and on-line programming. It allows ladder diagrams to be stored to floppy disk, uploaded/downloaded from/to the PLC. The software allows monitoring the operation of the ladder while it is executing. The software allows forcing system inputs/outputs (I/O) on and off. This is extremely valuable for troubleshooting. It also allows the programmer to document the ladder. This documentation is invaluable for understanding and troubleshooting ladder diagrams. The programmer can add notes, names of input or output devices, and comments that may be useful for troubleshooting and maintenance. The addition of notes and comments allows any technician to readily understand the ladder diagram. This allows any technician to troubleshoot the system, not just the person who developed it. The notes and/or comments could even specify replacement part numbers if so desired. This would facilitate rapid repair of any problems due to faulty parts.

The old way was that the person who developed the system had great job security because no one else could understand what had been done. A properly documented ladder allows any technician to understand it.

All of the leading brands of PLCs have software available so that a microcomputer can be used as the programming device. Square D calls their programming software "SY/MATE." Omron calls their programming software "SYSMATE."

The IEC has developed a standard for PLC programming. Giddings & Lewis has programming software available for their PLCs that meet the IEC world standard. The Giddings & Lewis software is called PICPro™. All of these software packages have the ability to program off-line and on-line, upload/download, comment, force I/O on and off, and document ladder logic. There is a lot of similarity between systems. Once the technician becomes familiar with one system, he/she can readily learn new systems.

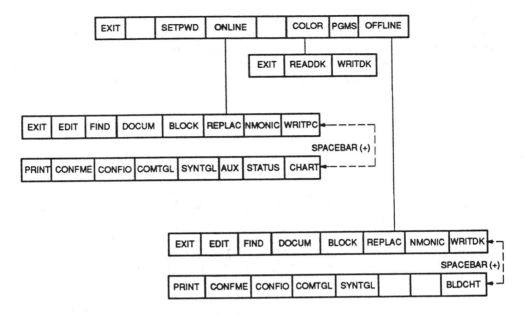

Figure 1-14. Menu tree for typical PLC programming software. This menu is from Siemens Industrial Automation, Inc. TISOFT. Each command represents a function key. When the function key is pressed, that function is accessed. For example, if the user chooses on-line another menu is presented. The user then has new choices, such as exit, edit (edit the program), find (this would search for certain parts of the program), document (this is used to document the ladder), etc. If the spacebar is pressed, the user is presented with eight other choices. If the off-line function key had been chosen, the user would have had many of the same options except that it is done in the computer memory and does not affect the PLC. Courtesy Siemens Industrial Automation, Inc..

Power Supply

The power supply is used to supply power for the central processing unit. Most PLCs operate on 115 Vac. This means that the input voltage to the power supply is 115 Vac. The power supply provides various dc voltages for the PLC components and CPU. On some PLCs the power supply is a separate module. This is usually the case when extra racks are used. Each rack must have its own power supply.

The user must determine how much current will be drawn by the I/O modules to ensure that the power supply will supply adequate current. Different types of modules draw different amounts of current. <u>Note</u>: This power supply is not typically used to power external inputs or outputs. The user must provide separate power supplies to power the inputs and outputs of the PLC. Some of the smaller PLCs do supply voltage to be used to power the inputs, however.

Input Section

The input portion of the PLC performs two vital tasks. It takes inputs from the outside world and protects the CPU from the outside world. Inputs can be almost any device. The input module converts the real-world logic level to the logic level required by the CPU. For example, a 250-Vac input module would convert a 250-Vac input to a low level dc signal for the CPU. Common input devices would include switches, sensors, etc. Other smart devices, such as robots, computers, and even PLCs, can act as inputs to the PLC.

The inputs are provided through the use of input modules. The user simply chooses input modules that will meet the needs of the application. These modules are installed in the PLC rack (see Figure 1-15). The PLC rack serves several functions. It is used to physically hold the CPU, power supply, and I/O modules. The rack also provides the electrical connections and communications between the modules, power supply, and CPU through the backplane.

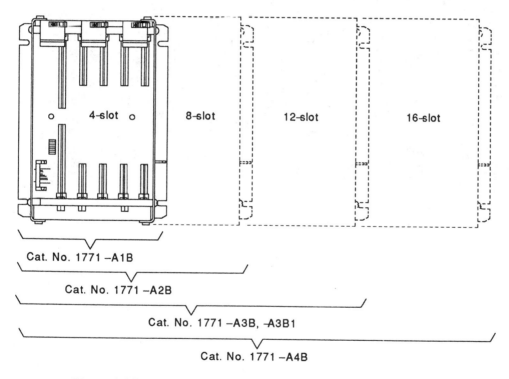

Figure 1-15. Various rack sizes for an Allen Bradley PLC. Courtesy of Allen Bradley Company, Inc., a unit of Rockwell International.

The modules are plugged into slots on the rack (see Figure 1-16). This ability to plug modules in and out easily is one of the reasons PLCs are so popular. The ability to change modules quickly allows very rapid maintenance and repair.

Input/output numbering is a result of which slot the module is plugged into. Omron calls their slots, channels. The combination of the channel number and actual input or output point on the module determine the number (see Figure 1-17).

Figure 1-16. Rack filled with modules. Courtesy of Siemens Industrial Automation, Inc..

I/O Channel # + Bit # (0-15)			I/O Address
Channel 150, bit 3	150	03	15003
Channel 1, bit 12	001	12	00112

Figure 1-17. How I/O is numbered in the Omron C200H PLC. Two examples are shown. If bit number 3 needs to be used on channel number 150, the I/O number would be 15003. If bit number 12 needs to be used on channel 1, the I/O number would be 00112.

There are cases when it is necessary to have more than one rack. Some medium-sized and large PLCs allow more than one rack. There are applications that have more I/O points than one rack can handle. There are also some cases when it is desirable to locate some of the I/O away from the PLC. For example, imagine a very large machine. Rather than run wires from every input and output to the PLC, an extra rack might be used. The I/O on one end of the machine is wired to modules in the remote rack. The I/O on the other end of the machine are wired to the main rack. The two racks communicate with one set of wiring rather than running all of the wiring from one end to the other.

Many PLCs allow the use of multiple racks (see Figure 1-18). When more than one rack is used it is necessary to identify which rack the I/O is in. Omron handles this through the channel number. Allen Bradley requires that the I/O number include the rack number (see Figures 1-19 to 1-22).

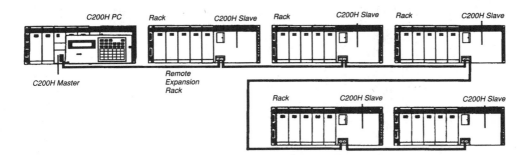

Figure 1-18. Use of multiple racks. In this case five additional racks were connected to the main PLC. This allows more I/O to be used. It also allows for remote mounting of the racks. Courtesy of Omron Electronics.

$$\begin{array}{c} 112 \\ \dashv\ \vdash \\ 07 \end{array}$$

Figure 1-19. Numbering of an Allen Bradley input. This input would correspond to input terminal 7 of group 2 of rack 1. The first 1 means that it is an input. Courtesy of Allen Bradley Company, Inc., a unit of Rockwell International.

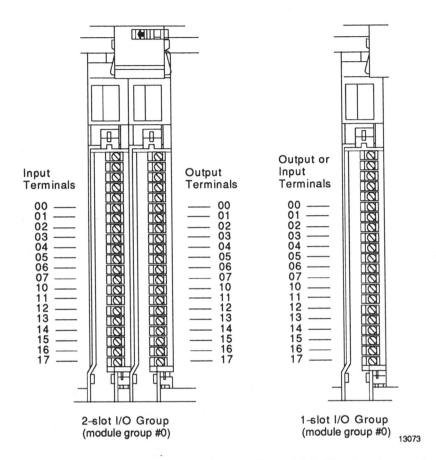

Input Terminals		Output Terminals	Output or Input Terminals	
00		00	00	
01		01	01	
02		02	02	
03		03	03	
04		04	04	
05		05	05	
06		06	06	
07		07	07	
10		10	10	
11		11	11	
12		12	12	
13		13	13	
14		14	14	
15		15	15	
16		16	16	
17		17	17	

2-slot I/O Group
(module group #0)

1-slot I/O Group
(module group #0)

13073

Figure 1-20. I/O group for an Allen Bradley PLC-5. The drawing on the left shows two 16-point modules which comprise one group. They are in module group 0 in this case. One word will be used for the 16 inputs and one word will be used for the 16 outputs. Note the octal numbering. The drawing on the right shows a one-slot I/O group (module group 0). An I/O group can contain up to 16 input terminals and 16 output terminals and can occupy 1/2, one, or two slots. Courtesy of Allen Bradley Company, Inc., a unit of Rockwell International.

Optical Isolation

The other task that the input section of a PLC performs is isolation. The PLC CPU must be protected from the outside world and at the same time be able to take input data from the outside world. This is typically done by opto-isolation (see Figure 1-23). "Opto-isolation" is short for "optical isolation." This means that there is no electrical connection between the outside world and the CPU. The two are separated optically. The outside world supplies a signal which turns on a light in the input card. The light shines on a receiver and the receiver turns on. There is no connection between the two.

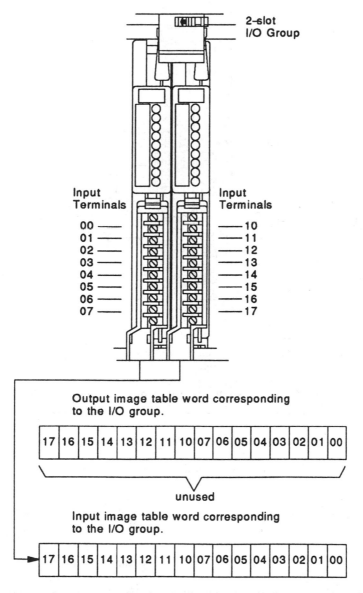

Figure 1-21. Two 8-point input modules using one word of the input image table for a PLC-5. Note that the inputs are numbered 00 to 17. Remember that octal numbering is used. This is 16 inputs or one word. Note that two slots are used to form this one I/O group. Courtesy of Allen Bradley Company, Inc., a unit of Rockwell International.

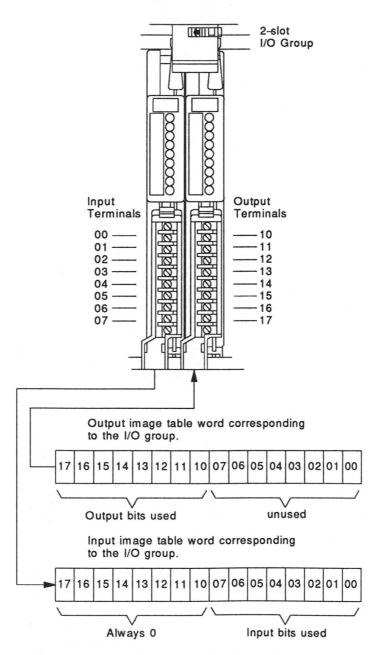

Figure 1-22. Use of one 8-point input module and one 8-point output PLC-5 module. In this case two slots are used to create one group. Because input and output modules have both been used in this group, two words of memory must be used. The first eight bits of the input image table word are used for inputs 00-07 and the last eight bits of the word from the output image table are used for outputs 10 to 17. If two 16-point modules are used, one must be an input module and one must be an output module. In that case both modules use a full word of memory. Courtesy of Allen Bradley Company, Inc., a unit of Rockwell International.

The light separates the CPU from the outside world up to very high voltages. Even if there was a large surge of electricity, the CPU would be safe. (Of course, if the voltage is too large, the opto-isolator can fail and may cause a circuit failure.) Optical isolation is used for inputs and outputs.

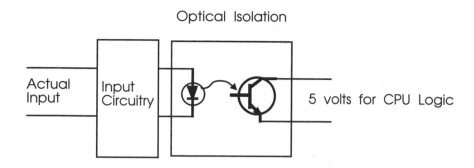

Figure 1-23. Typical optical isolation circuit. Note: the arrow represents the fact that only light travels from between the input circuitry and the CPU circuitry. There is no electrical connection.

Input modules provide the user with various troubleshooting aids. There are normally light emitting diodes (LEDs) for each input. If the input is on, the CPU should see the input as a high (or a 1).

Input modules also provide circuits that debounce the input signal. Many input devices are mechanical and have contacts. When these devices close or open there are unwanted "bounces" that close and open the contacts. Debounce circuits make sure that the CPU sees only debounced signals. The debounce circuit also helps eliminate the possibility of electrical noise from firing the inputs.

Inputs to Input Modules

Sensors are commonly used as inputs to PLCs. Sensors can be purchased for a variety of purposes. They can sense part presence, count pieces, measure temperature, pressure, or size, sense for proper packaging, and so on.

There are also sensors that are able to sense any type of material. Inductive sensors can sense ferrous metal objects, capacitive sensors can sense almost any material, and optical sensors can detect any type of material.

Other devices can also act as inputs to a PLC. Smart devices such as robots, computers, vision systems, etc., often have the ability to send signals to a PLC's input modules (see Figure 1-24). This can be used for handshaking during operation. A robot, for example, can send the PLC an input when it has finished a program.

Sensor types and use are covered in detail in Chapter 5.

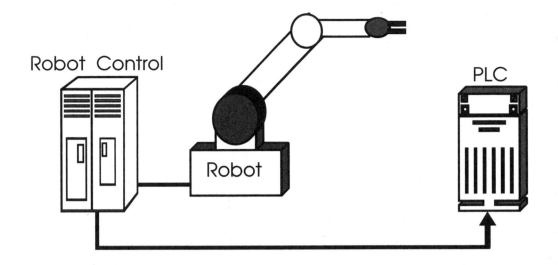

Robot Control

PLC

Robot

*Figure 1-24. A "smart" device can also act as an input device to a PLC.
The PLC can also output to the robot. Robots and some other devices
typically have a few digital outputs and inputs available for this purpose. The
use of these inputs and outputs allows for some basic handshaking between
devices. Handshaking means that the devices give each other permission to
perform tasks at some times during execution to assure proper performance
and safety. When devices communicate with a digital signal it is called
primitive communication.*

Output Section

The output section of the PLC provides the connection to real-world output devices. The output
devices might be motor starters, lights, coils, valves, etc. Output modules can be purchased to
handle dc or ac voltages. They can be used to output analog or digital signals.

A digital output module acts like a switch. The output is either energized or deenergized. If the
output is energized, the output is turned on, just like a switch.

The analog output module is used to output an analog signal. An example of this is a motor
whose velocity we would like to control. An analog module puts out a voltage that corresponds
to the desired speed.

Output modules can be purchased with various output configurations. They are available as
modules with 8, 16, and 32 outputs. Modules with more than eight outputs are sometimes called
high-density modules. They are generally the same size as the eight-output modules. They just
have many more components within the module. For that reason high-density modules will not
handle as much current for each output. This is because of the size of components and the heat
generated by them.

Current Ratings

Module specifications will list an overall current rating and an output current rating. For example, the specification may give each output a current limit of 1 ampere. If there are eight outputs, we would assume that the output module overall rating would be 8A. This is a poor leap in logic. The overall rating of the module current will normally be less than the total of the individuals. The overall rating might be 6A. The user must take this into consideration when planning the system. Normally, each of the eight devices would not pull their 1A at the same time.

Output Image Table

The output image table is a part of CPU memory (see Figure 1-25). The users logic determines whether an output should be on or off. The CPU evaluates the user's ladder logic. If it determines that an output should be on it stores a one in the bit that corresponds to that output. The one in the output image table is used to turn the actual output through an isolation circuit.

The outputs of small PLCs are often relays. This allows the user to mix and match output voltages or types. For example, some outputs could then be ac and some dc. There are relay output modules available for some of the larger PLCs too. The other choices are transistors for dc outputs and triacs for ac outputs.

Output modules are covered in greater detail in Chapter 6.

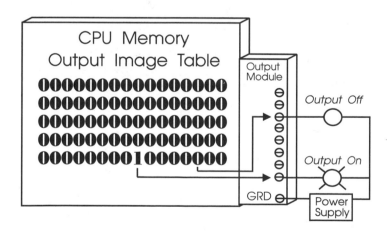

Figure 1-25. How a typical PLC handles outputs. The CPU memory contains a section called the output image table. The output image table contains the desired states of all outputs. If there is a 1 in the bit that corresponds to the output, the output is turned on. If there is a 0, the output is turned off. This is called active-high logic.

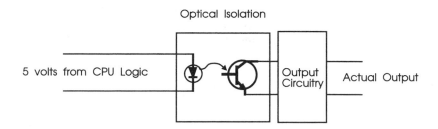

Optical Isolation

5 volts from CPU Logic

Output Circuitry

Actual Output

Figure 1-26. How PLC output isolation works. The CPU logic provides a five volt signal which turns on the LED. The light from the LED is used to fire the base of the output transistor. There is no electrical connection between the CPU and the outside world. The output circuitry normally provides fused protection for each output. The output circuitry also contains transistors, triacs, or relays to handle the output current.

Figure 1-27. A few possible output devices including a contactor, starter, ac motor, and a valve.

PLC Applications

Programmable logic controllers are being used for a wide variety of applications (see Figures 1-28 to 1-30).

They are used to replace hardwired logic in older machines. This can reduce the downtime and maintenance of older equipment. More importantly PLCs can increase the speed and capability of older equipment. Retrofitting an older piece of equipment with a PLC for control is almost like getting a new machine.

PLCs are being used to control such processes as chemical production, paper production, steel production, food processing, and so on. In processes such as these they are used to control temperature, pressure, mixture, concentration, etc. They are used to control position and velocity in many kinds of production processes. For example, they can be used to control complex automated storage and retrieval systems. They can also be used to control equipment such as robots and production machining equipment.

Figure 1-28. This figure shows a PLC-controlled injection molding machine. Photo courtesy of Allen Bradley Company, Inc., a unit of Rockwell International.

Many small companies have started up recently to produce special-purpose equipment. This equipment is normally controlled by PLCs. It is very cost effective for these companies to use PLCs. Examples are conveyors and palletizing, packaging, processing, material handling, etc. Without PLC technology many small equipment design companies might not exist.

Figure 1-29. Large flexographic printing press. It is controlled by a PLC. Flexographic presses are used to produce packaging materials and other types of printed materials.

Figure 1-30. Food packaging line controlled by a Square D PLC. Courtesy Square D.

Chapter 1: Overview of Programmable Logic Controllers

PLCs are being used extensively in position and velocity control. A PLC can control position and velocity much more quickly and accurately than can mechanical devices such as gears, cams, etc. An electronic system of control is not only faster, but does not wear out and lose accuracy as do mechanical devices.

PLCs are used for almost any process one can think of. There are companies that have PLC equipped railroad cars that regrind and true the rail track as they travel. They have been used to ring the perfect sequences of bells in church bell towers, at the exact times during the day and week. PLCs are used in lumber mills to grade, size, and cut lumber for optimal output.

The uses of PLCs are limited only by the imagination of the engineers and technicians who use them.

Questions

1. The PLC was developed to:

    a. make manufacturing more flexible.

    b. be easy for electricians to work with.

    c. make systems more reliable.

    d. all of the above.

2. The PLC is programmed by technicians using:

    a. the "C" programming language.

    b. ladder logic.

    c. the choice of language used depends on the manufacturer.

    d. none of the above.

3. The most common programming device for PLCs is the:

    a. dumb terminal.

    b. dedicated programming terminal.

    c. handheld programmer.

    d. personal computer.

4. CPU stands for central processing unit. True or false?

5. Opto-isolation:

    a. is used to protect the CPU from real-world inputs.

    b. opto-isolation is not used in PLCs. Isolation must be provided by the user.

    c. is used to protect the CPU from real-world outputs.

    d. both a and c.

6. EPROM is:

    a. electrically erasable memory.

    b. electrically programmable RAM.

    c. erased by exposing it to ultraviolet light.

    d. programmable.

    e. both c and d.

7. RAM typically holds the operating system. True or false?

8. Typical program storage devices for ladder diagrams include:

   a. computer disks.

   b. EEPROM.

   c. static RAM cards.

   d. all of the above.

   e. none of the above.

9. Input devices would include the following:

   a. switches.

   b. sensors.

   c. other smart devices.

   d. all of the above.

   e. none of the above.

10. Output modules can be purchased with which of the following output devices:

    a. Transistor outputs.

    b. Triac outputs.

    c. Relay outputs

    d. All of the above are correct.

    e. Both a and c.

11. List at least four major advantages of the use of a microcomputer for programming PLCs.

12. What is a memory map? (Make sure that you describe the types of information that can be found on a memory map.)

13. If an output modules current rating is 1 A per output and there are eight outputs the current rating for the module is 8 A. True or false? Explain your answer.

14. List at least four reasons why PLCs have replaced hardwired logic in industry.

15. Draw a block diagram of a typical PLC that shows the major components.

16. Describe how the status of real-world inputs are stored in the PLC memory.

17. Describe how the status of real-world outputs are stored in PLC memory.

18. Define the term <u>debounce</u>. Why is it so important?

19. What is the difference between on-line and off-line programming?

20. What does it mean to force I/O?

Chapter 2

Overview of Number Systems

A knowledge of various numbering systems is essential to the use of PLCs. In addition to decimal, binary, octal and hexadecimal are regularly used. An understanding of the systems will make the task of working with PLCs an easier one. In this chapter we examine each of these systems.

Objectives

Upon completion of this chapter, the student will be able to:

Explain each of the numbering systems.

Explain the benefits of typical number systems and why each is used.

Convert from one number system to another.

Explain how input or output modules might be numbered using the octal or hexadecimal number system.

Use each number system properly.

Explain such terminology as <u>most significant</u>, <u>least significant</u>, <u>nibble</u>, <u>byte</u>, and <u>word</u>.

Decimal

A short review of the basics of the decimal system will help in a thorough understanding of the other number systems. Although calculators will do the tedious work of number conversion between systems, it is vital that the technician be comfortable with the number systems. Binary, octal, and hexadecimal are regularly used to identify inputs/outputs, memory addresses etc. The technician who is comfortable in understanding and using these systems will have an easier time with PLCs.

The Decimal number system uses 10 digits: zero through nine. The highest digit is nine in the decimal system. Zero through nine are the only digits allowed.

Decimal System					
100,000s	10,000s	1000s	100s	10s	1s

Figure 2-1. Weights of the decimal system.

The first column in decimal can be used to count up to nine items. Another column must be added if the number is larger than nine. The second column can also use the digits zero through nine. This column is weighted, however. The second column is used to tell how many 10s there are. For example, the number 23 represents two 10s and three 1s. By using one column we were able to count to nine. If we use two columns we can count to 99. The first column can hold up to nine. The second column can hold up to 90 (9 tens), exactly ten times as much as the first column. In fact, in the decimal system, each column is worth ten times as much as the preceding column. The third column represents the number of hundreds (10 times 10). For example, 227 would represent two 100s, two 10s, and seven 1s. (see Figure 2-2).

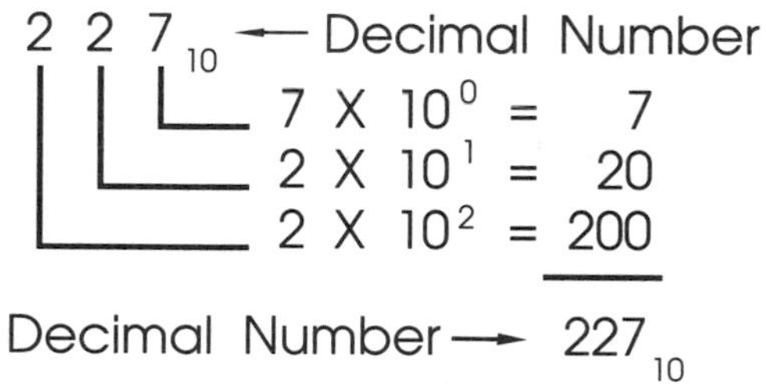

$$2\ 2\ 7_{10} \longleftarrow \text{Decimal Number}$$

$$7 \times 10^0 = 7$$
$$2 \times 10^1 = 20$$
$$2 \times 10^2 = 200$$

$$\text{Decimal Number} \longrightarrow 227_{10}$$

Figure 2-2. Relationship between the weights of each column and the decimal number 227.

The decimal system is certainly the simplest for us because we have used it all of our lives. The other systems are based on the same principles as the decimal system. It would certainly be easier for us if there were only one system, but the computer cannot "think" in decimal. The computer can only work with binary numbers. In fact, the other number systems are very convenient for certain uses and actually simplify some tasks.

Binary Numbering System

The binary numbering system is based on only two digits: zero and one. A computer is a digital device. It works with voltages, on or off. Computer memory is a series of zeros and ones.

The binary system works just like decimal. The first column holds the number of 1s (see Figure 2-3). Since the only possible digits in the first column are zero or one, it should be clear that the first column can hold zero 1s or one 1. Thus we can only count up to 1 using the first column in binary.

The second column holds the number of 2s. There can be zero 2s or one 2. The binary number 10 would equal one 2 plus zero 1s. The number 10 in binary is two in the decimal system (one 2 + zero 1s). The binary number 11 would be 3 in decimal (one 2 + one 1) (see Figure 2-4).

The third column is the number of 4s. Thus binary 100 would be equal to decimal 4.

The fourth column is the number of 8s, the fifth column is the number of 16s, the sixth column is the number of 32s, the seventh is the number of 64s, and the eighth column is the number of 128s. As you can see, each column's value is twice as large as that of the previous column.

Binary System					
32s	16s	8s	4s	2s	1s

Figure 2-3. Weights of each column in the binary system. The first column is the number of 1s, the second is the number of 2s, and so on.

Binary				Decimal
8s	4s	2s	1s	
0	0	0	0	0
0	0	0	1	1
0	0	1	0	2
0	0	1	1	3
0	1	0	0	4
0	1	0	1	5
0	1	1	0	6
0	1	1	1	7
1	0	0	0	8
1	0	0	1	9
1	0	1	0	10
1	0	1	1	11
1	1	0	0	12
1	1	0	1	13
1	1	1	0	14
1	1	1	1	15

Figure 2-4. Comparison of the binary and decimal numbers from 0 through 15.

The value of the column in binary can be found by raising 2 to the power represented by that column. For example, the third column's weight could be found by raising 2 to the second power (2 X 2 = 4). Remember that the first column is column number 0, the second is column number 1, and so on. The weight of the fourth column is eight (2 X 2 X 2) (see Figure 2-5).

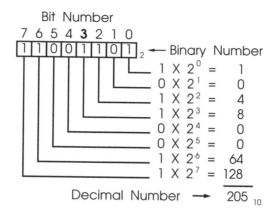

$$1 \times 2^0 = 1$$
$$0 \times 2^1 = 0$$
$$1 \times 2^2 = 4$$
$$1 \times 2^3 = 8$$
$$0 \times 2^4 = 0$$
$$0 \times 2^5 = 0$$
$$1 \times 2^6 = 64$$
$$1 \times 2^7 = 128$$

Decimal Number → 205_{10}

Figure 2-5. Relationship between binary and decimal. The binary number 11001101 is equal to 205 decimal.

Binary is used extensively because it is the only numbering system that a computer can deal with. It is also quite useful when considering digital logic, as a 1 can represent one state and a zero the opposite state. For example, a light is either on or off. See Figure 2-6 for the appearance of a binary word.

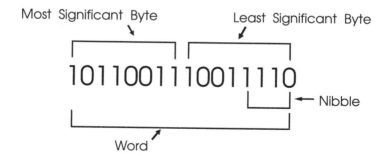

Figure 2-6. A 16-bit binary number. The bit on the right is the least significant bit. The bit on the left is the most significant bit. The next unit of grouping is the nibble. A nibble is 4 bits. The next grouping of a binary number is the byte. A byte is 8 bits. Note that the first 8 digits are called the least significant byte and the last 8 bits are called the most significant. The next grouping is called the "word." The size of a word is dependent on the processor. A 16-bit processor has a 16-bit word. A 32-bit processor has a 32-bit word.

Binary Coded Decimal System

Binary-coded decimal (BCD) involves the blending of the binary and decimal systems. In BCD 4 binary bits are used to represent one decimal digit. These 4 bits are used to represent the numbers zero through nine. Thus 0111 binary would be 7 decimal. (See Figure 2-7.) The difference in BCD is in the way numbers above decimal nine are represented. For example, the decimal number 43 would be 0100 0011 in BCD. The first 4 bits (least significant bits) represent the decimal 3. The second 4 bits (most significant bits) represent the decimal 4. In BCD the first 4 bits represent the number of 1s in a decimal number, the second 4 represent the number of 10s, the third four represent the number of 100s, and so on.

The BCD format is very popular for output from instruments. Measuring devices will typically output BCD values. Some input devices use the BCD system to output their value. Thumbwheels are one example. A person dials in a decimal digit between zero and nine. The thumbwheel outputs 4 bits of data. The 4 bits are BCD. For example, if an operator were to dial in the number eight, the output from the BCD thumbwheel would be 1000. PLCs can easily accept BCD input.

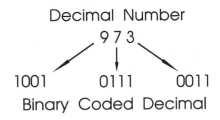

Figure 2-7. How the decimal number 973 would be represented in the binary-coded decimal (BCD) system. Each decimal digit is represented by its four-digit binary number. _Caution:_ BCD is not the same as binary. The decimal number 973 is 1001 0111 0011 in BCD and is 0011 1100 1101 in binary.

Octal

The octal system is based on the same principles as binary and decimal except that it is base eight. There are eight possible digits in the octal system (0, 1, 2, 3, 4, 5, 6, and 7). The first column in an octal number is the number of 1s. The second column is the number of 8s, the third column is the number of 64s, the fourth column is the number of 512s, the fifth column is the number of 4096s, and so on.

The weights of the columns can be found by using the same method as that used for binary. The number eight is simply raised to the power represented by that column. The first column (column 0) represents eight to the zero power (1 by definition). Remember that the first column is column zero. The weight of the second column (column one) is found by raising eight to the first power (8X1). The weight of the third column (column two) is found by raising eight to the second power (8X8) (see Figure 2-8).

Octal System

32,768s	4,096s	512s	64s	8s	1s

Figure 2-8. Weights of the columns in the octal number system. The first column is the number of 1s, the second the number of 8s, and so on.

The actual digits in the octal number system are 1, 2, 3, 4, 5, 6, and 7. If we must count above seven we must use the next column. For example, lets count to ten in octal: 1, 2, 3, 4, 5, 6, 7, 10, 11, and 12. The 12 represents one 8 and two 1s (8+2=10).

The number 23 decimal would be 27 in octal. Two 8s and seven 1s is equal to 23. The number 3207 octal would be 1671 in decimal. Figure 2-9 shows how an octal number can be converted to a decimal number.

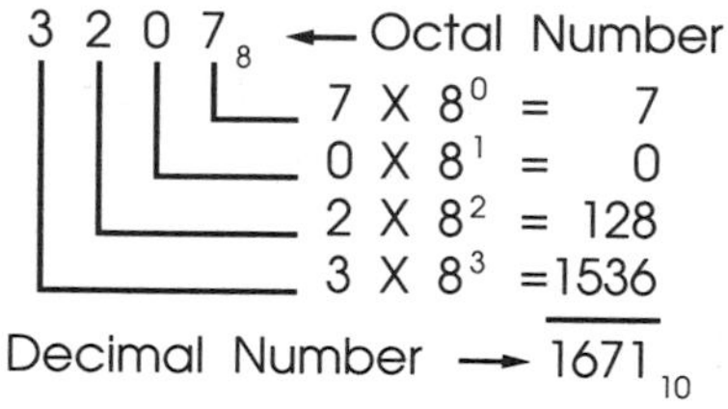

$$
\begin{array}{l}
3\ 2\ 0\ 7_8 \leftarrow \text{Octal Number} \\
\quad\quad\quad 7 \times 8^0 = \quad 7 \\
\quad\quad\quad 0 \times 8^1 = \quad 0 \\
\quad\quad\quad 2 \times 8^2 = 128 \\
\quad\quad\quad 3 \times 8^3 = 1536 \\
\text{Decimal Number} \rightarrow 1671_{10}
\end{array}
$$

Figure 2-9. How the octal number 3207 is converted to the decimal number 1671.

Some PLC manufacturers use octal to number input and output modules, and also to number memory addresses. For example (assume the use of input cards with eight inputs per card), the first eight inputs on the first card would be numbered 0, 1, 2, 3, 4, 5, 6, and 7. The next input card numbering would begin with octal 10, 11, 12, 13, 14, 15, 16, and 17 (see Figure 2-10).

This makes it very easy to find the location of inputs or outputs. The least significant digit is used to specify the actual input/output number, and the most significant digit is used to specify the particular card where the input/output is located.

Octal is also used by some manufacturers for numbering memory. For example, Siemens Industrial Automation, Inc. has 128 timers and counters available for the 405 series of PLCs. The timers are numbered 0 to 177 octal. This equates to 0 to 127 decimal.

Hexadecimal

Hexadecimal normally causes the most trouble for people. Hexadecimal is based on the same principles as the other numbering systems we examined. Hexadecimal has 16 possible digits. Hexadecimal has an unusual twist, however. Hexadecimal uses numbers and also the letters A to F. This is a little confusing at first.

Input Module 0	Input Module 1	Input Module 2	Input Module 3
I00	I10	I20	I30
I01	I11	I21	I31
I02	I12	I22	I32
I03	I13	I23	I33
I04	I14	I24	I34
I05	I15	I25	I35
I06	I16	I26	I36
I07	I17	I27	I37

Figure 2-10. How some manufacturers number input/output (I/O) modules using the octal numbering system. The first input module is numbered 0. The first input on the module would be called input 00 (the first zero represents the first module, the second zero means that it is the first input.). The eighth input on the first module would be called 07. The eighth input on the fourth module would be called input 37. (Remember that the first module is zero; the fourth module would be number three.) The first output module would begin numbering as output module 0 and each output would be numbered just like the input modules were. For example, the fifth output on the second output module would be numbered output 14. Siemens Industrial Automation, Inc. uses this numbering scheme with their 405 series.

In hexadecimal (hex for short) we count 0, 1, 2, 3, 4, 5, 6, 7, 8, and 9. After the number nine the counting changes: ten becomes A, eleven is B, twelve is C, thirteen is D, fourteen is E, and fifteen is F. (See Figure 2-11).

The first column (column 0) in hex is the number of 1s (see Figure 2-12). The second column is the number of 16s. The third column is the number of 256s, and so on.

The weights can be found in the same manner as in the binary system. In the hexadecimal system sixteen would be raised to the power of the column. For example, the weight of the third column (column 2) would be 16 to the second power (16X16=256).

Figure 2-13 shows how a hexadecimal number can be converted to a decimal number.

Hexadecimal	Decimal
0	0
1	1
2	2
3	3
4	4
5	5
6	6
7	7
8	8
9	9
A	10
B	11
C	12
D	13
E	14
F	15
10	16
11	17
12	18
13	19
14	20

Figure 2-11. Comparison of the hexadecimal and decimal systems.

Hexadecimal System					
1,048,576s	65,536s	4,096s	256s	16s	1s

Figure 2-12. Weights of the columns in the hexadecimal system. The first column is the number of 1s , the second column is the number of 16s, and so on.

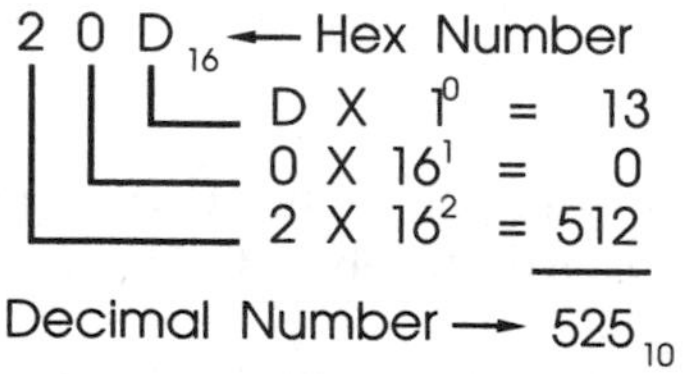

$$2\ 0\ D_{16} \leftarrow \text{Hex Number}$$

$$D \times 1^0 = 13$$
$$0 \times 16^1 = 0$$
$$2 \times 16^2 = 512$$

$$\text{Decimal Number} \rightarrow 525_{10}$$

Figure 2-13. Weights of the hexadecimal number system converted to the decimal weighting system. The hex number 20D is equal to the decimal number 525.

Hexadecimal numbers are easier (less cumbersome) for us to work with than binary numbers. It is easy to convert between the two systems (see Figures 2-14 and 2-15). Each hex digit is simply converted to its four-digit binary equivalent. The result is the binary equivalent of the whole hex number.

The same process works in reverse. Any binary number can be converted to its hex equivalent by breaking the binary number into four-digit pieces and converting each four-bit piece to its hex equivalent. The resulting numbers are equal in value.

For an overview of the four number systems discussed, see Figure 2-16.

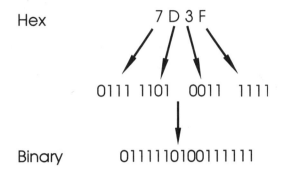

Figure 2-14. This figure shows the conversion of a hexadecimal number to its binary equivalent. Each hex digit is simply converted to its four-digit binary. The result is a binary equivalent. In this case hex 7D3F is equal to binary 0111110100111111. (Which number would you rather work with?)

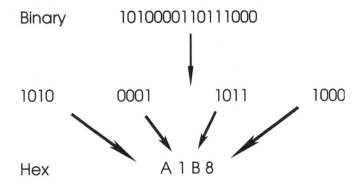

Figure 2-15. Conversion of a binary number to a hex number. The binary number is broken into four bit pieces (4 bits are called a nibble - half a byte) and each 4-bit nibble is converted to its hex equivalent. The binary number 1010000110111000 is equal to hex A1B8. (Which would you prefer to work with?)

Hexadecimal	Decimal	Octal 8s	Octal 1s	Binary 8s	Binary 4s	Binary 2s	Binary 1s
0	0	0	0	0	0	0	0
1	1	0	1	0	0	0	1
2	2	0	2	0	0	1	0
3	3	0	3	0	0	1	1
4	4	0	4	0	1	0	0
5	5	0	5	0	1	0	1
6	6	0	6	0	1	1	0
7	7	0	7	0	1	1	1
8	8	1	0	1	0	0	0
9	9	1	1	1	0	0	1
A	10	1	2	1	0	1	0
B	11	1	3	1	0	1	1
C	12	1	4	1	1	0	0
D	13	1	5	1	1	0	1
E	14	1	6	1	1	1	0
F	15	1	7	1	1	1	1

Figure 2-16. This figure shows a comparison of four number systems for the numbers 0 to 15.

Questions

1. Complete the following table.

	Binary	Octal	Decimal	Hexadecimal
a.	101			5
b.		11		
c.			15	
d.				D
e.		16		
f.	1001011			
g.		47		
h.			73	

2. Number the following module input/output modules using the octal method.

	Input Module 0	Input Module 1	*Output Module 0*	Input Module 2
A.				
B.				
C.				
D.				
E.				
F.				
G.				
H.				

3. Define each of the following:

   a. bit

   b. nibble

   c. byte

   d. word

4. Why is the binary number system used in computer systems?

5. Complete the following table.

	Binary	Hexadecimal
a.	1011001011111101	
b.		1A07
c.	100100000111	
d.		C17F
e.	0010001111000010	
f.		D91C
g.	0011010111100110	
h.		ECA9
i.	0101001011000101	

6. True or False. A word is sixteen bits. Explain your answer.

7. True or False. One "K" of memory is exactly one thousand bytes.

8. How can the weight of the fifth column of a binary number be calculated?

9. How can the weight of the fourth column of a hex number be calculated?

10. What is the BCD system normally used for?

Chapter 3

Fundamentals of Programming

In this chapter we examine the basics of ladder logic programming. Terminology and common symbols are emphasized. The student will learn how to write basic ladder logic programs. Contacts, coils, timers, and counters are used.

Objectives

Upon completion of this chapter, the student will be able to:

Describe the basic process of writing ladder logic.

Define such terms as, <u>contact</u>, <u>coil</u>, <u>rung</u>, <u>scan</u>, <u>normally open</u>, <u>normally closed</u>, and <u>timer</u>.

Write ladder logic for simple applications.

Ladder Logic

Programmable controllers are primarily programmed in ladder logic. Ladder logic is really just a symbolic representation of an electrical circuit. Symbols were chosen that actually looked similar to schematic symbols of electrical devices. This made it easy for the plant electrician to learn to use the PLC. An electrician who has never seen a PLC can understand a ladder diagram.

The main function of the PLC program is to control outputs based on the condition of inputs. The symbols used in ladder logic programming can be divided into two broad categories: contacts (inputs) and coils (outputs).

Contacts

Most inputs to a PLC are simple devices that are either on or off. These inputs are sensors and switches that detect part presence, empty or full, and so on. The two common symbols for contacts are shown in Figure 3-1.

Normally Open Contact   **Normally Closed Contact**

Figure 3-1. A normally open and a normally closed contact.

Contacts can be thought of as switches. There are two basic kinds of switches, normally open and normally closed. A normally open switch will not pass current until pressed. A normally closed switch will allow current flow until it is pressed. Think of a doorbell switch. Would you use a normally open switch or a normally closed switch for a doorbell?

If you chose the normally closed switch, the bell would be on continuously until someone pushes the switch. Pushing the switch opens the contacts and stops current flow to the bell. The normally open switch is the necessary choice. If the normally open switch is used, the bell will not sound until someone pushes the button on the switch.

Sensors are used to sense for the presence of physical objects or quantities. For example, one type of sensor might be used to sense when a box moves down a conveyor, and a different type might be used to measure a quantity such as heat. Most sensors are switch-like. They are on or off depending on what the sensor is sensing. Like switches, sensors can be purchased that are either normally open or normally closed.

Imagine, for example, a sensor that is designed to sense a metal part as the part passes the sensor. We could buy a normally open or a normally closed sensor for the application. If it were desired to notify the PLC every time a part passed the sensor, a normally open sensor might be chosen. The sensor would turn on only if a metal part passed in front of the sensor. The sensor would turn off again when the part was gone.

The PLC could then count the number of times the normally open switch turned on (closed) and would know how many parts had passed the sensor. Normally closed sensors and switches are often used when safety is a concern. These are examined later.

Coils

Contacts are input symbols, coils are output symbols. Outputs can take various forms: motors, lights, pumps, counters, timers, relays, and so on.

The basic coil is shown in Figure 3-2.

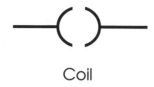

Coil

Figure 3-2. The ladder logic symbol for a coil.

A coil is simply an output. The PLC examines the contacts (inputs) in the ladder and turns the coils (outputs) on or off depending on the condition of the inputs.

Ladder Diagrams

The basic ladder diagram looks similar to a ladder. There are two uprights and there are rungs that make up the PLC ladder. The left and right upright represent power. If we connect the left and right uprights, power could flow through the rung from the left upright to the right upright.

Figure 3-3. A very simple conceptual view of a ladder diagram.

Consider the doorbell example again, one input and one output. The ladder for the PLC would be only one rung. See Figure 3-3. The real-world switch would be connected to input number zero of the PLC. The bell would be connected to output number zero of the PLC (see Figure 3-4). The uprights represent a dc voltage that will be used to power the doorbell. If the real-world doorbell switch is pressed, power can flow through the switch to the doorbell.

The PLC would then run the ladder. The PLC will monitor the input continuously and control the output. This is called <u>scanning</u>. The amount of time it takes for the PLC to go through the ladder logic each time is called the <u>scan time</u>. Scan time varies from PLC to PLC. Most applications do not require extreme speed, so any PLC is fast enough; even a slow PLC scan time would be in milliseconds. The longer the ladder logic, the more time it takes to scan.

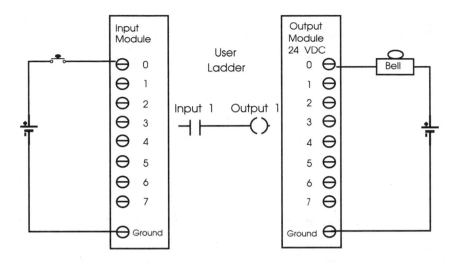

Figure 3-4. Conceptual view of a PLC system. The real-world inputs are attached to an input module (left side of the figure). Outputs are attached to an output module (right side of the figure). The center of the figure shows the logic that the CPU must evaluate. The CPU evaluates user logic by looking at the inputs and then turns on outputs based on the logic. In this case if input 0 (a normally open switch) is closed, output 0 (the doorbell) will turn on.

The scan cycle is illustrated by Figure 3-5. Note that this one rung of logic represents our entire ladder. Each time the PLC scans the doorbell ladder, it checks the state of the input switch <u>before</u> it enters the ladder (time period 1). While in the ladder, the PLC then decides if it needs to change the state of any outputs (evaluation during time period 2). <u>After</u> the PLC finishes evaluating the logic (time period 2), it turns on or off any outputs based on the evaluation (time period 3). The PLC then returns to the top of the ladder, checks the inputs again and repeats the entire process. It is the total of these three stages that make up scan time. We discuss the scan cycle more completely later in this chapter.

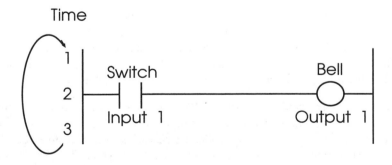

Figure 3-5. This figure shows an example of how a user's ladder logic is continually scanned.

Normally closed contacts

The second type of contact is the normally closed contact. The normally closed contact will pass power until it is activated. A normally closed contact in a ladder diagram will pass power while the real-world input associated with it is off.

A home security system is an example of the use of normally closed logic. Assume that the security system was intended to monitor the two entrance doors to a house. One way to wire the house would be to wire one normally open switch from each door to the alarm, just like a doorbell switch (Figure 3-6). Then if a door opened, it would close the switch and the alarm would sound. This would work, but there are problems.

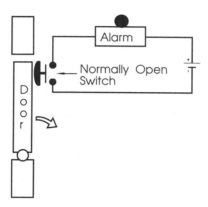

Figure 3-6. A conceptual diagram of a burglar alarm circuit. This is the wrong way to construct this type of application. In this case the homeowner would never know if the system failed. The correct method would be to use normally closed switches. The control system would then monitor the circuit continuously to see if the doors opened or a switch failed.

Assume that the switch fails. A wire might be cut accidentally, or a connection might become loose, or a switch breaks, etc. There are many ways in which the system could fail. The problem is that the homeowner would never know that the system was not working. An intruder could open the door, the switch would not work, and the alarm would not sound. Obviously, this is not a good way to design a system.

The system should be set up so that the alarm will sound for an intruder and will also sound if a component fails. The homeowner surely wants to know if the system fails. It is far better that the alarm sounds when the system fails and there is no intruder than not to sound if the system fails and there is an intruder.

Considerations such as these are even more important in an industrial setting where failure could cause an injury.

The procedure of programming to assure safety is called <u>fail-safe</u>. The programmer must carefully design the system and ladder logic so that if a failure occurs, people and process are safe. Consider Figure 3-7. If the gate is opened, it opens the normally closed switch. The PLC would see that the switch had opened and would sound an alarm immediately to protect whomever had entered the work cell. (In reality, we would sound an alarm and stop the robot to protect the intruder.)

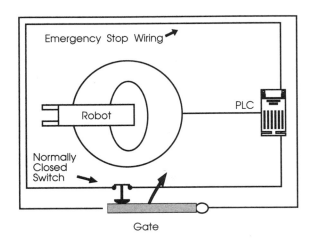

Figure 3-7. Cell application. This figure represents a robot cell. There is a fence around the cell with one gate. There is a PLC used as a cell controller. There is a safety switch to make sure no one enters the cell while the robot is running. If someone enters the cell the PLC will sense that the switch opened and sound an alarm. In this case a normally closed switch was used. If the wiring or switch fails the PLC will think someone entered the cell, and sound an alarm. This system would "fail-safe."

The normally closed switch as used in ladder logic can be confusing. The normally closed contact in our ladder passes electricity if the input switch is off. (The alarm would sound if the switch in the cell gate opened.) The switch in the gate of the cell is a normally closed switch. (The switch in the cell normally allows electricity to flow.) If someone opens the gate, the normally closed gate switch opens, stopping electrical flow (see Figure 3-8). The PLC sees that there is no flow, the normally closed contact in the ladder allows electricity to flow, and the alarm is turned on.

Assume that a tow motor drives too close to the cell and cuts the wire that connects the gate safety switch to the PLC. What will happen? The alarm will sound. Why? Because the wire being cut is similar to the gate opening the switch. Is it a good thing that the alarm sounds if the wire gets cut? Yes. This warns the operator that something failed in the cell. The operator could then call maintenance and have the cell repaired. This is "fail-safe," something in the cell failed and the system was shut down by the PLC so that no one would be hurt. The same would be true if the gate safety switch were to fail. The alarm would sound. If the switch were opened (someone opened the gate to the cell), the PLC would see that there is no power at the input (see Figure 3-9). The normally closed contact in the ladder logic is then closed, allowing electricity to flow. This causes the alarm to sound.

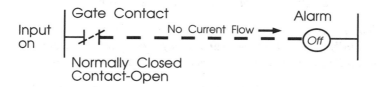

Figure 3-8. *This figure shows one rung of a ladder diagram. A normally closed contact is used in the ladder. If the switch associated with that contact is closed, it forces the normally closed contact open. No current flows to the output (the alarm). The alarm is off.*

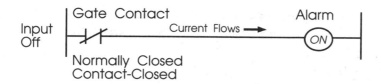

Figure 3-9. *In this figure the input is off. (Someone opened the gate and opened the switch.) The normally closed contact is true when the input is false, so the alarm sounds. The same thing would happen if a tow motor cut the wire that led to the safety sensor. The input would go low and the alarm would sound.*

Consider the rungs shown in Figure 3-10 and determine whether the output coils are on or off. (The answers follow.) Pay particular attention to the normally closed examples.

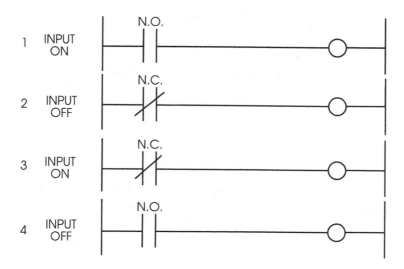

Figure 3-10.

Answers to the ladder logic in Figure 3-10 are shown below.

1. *The output in example 1 would be on. The input associated with the normally open contact (or examine-on) is one that would close the normally closed contact in the ladder and pass power to the output.*

2. *The output in example 2 would be on. The real-world input is off. The normally closed contact is closed because the input is off.*

3. *The output in example 3 is off. The real-world input is on which forces the normally closed contact open (or the rung is false because the examine-off input is on).*

4. *The output in example 4 would be off. The real world input is off, so the normally open contact remains open and the output off.*

Multiple Contacts

More than one contact can be put on the same rung. For example, think of a drill machine. The engineer wants the drill press to turn on only if there is a part present and the operator has one hand on each of the start switches (see Figures 3-11 and 3-12). This would ensure that the operator's hands could not be in the press while it is running.

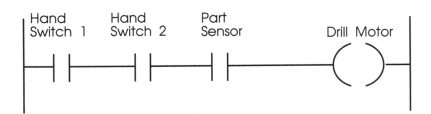

Figure 3-11. This figure shows a series circuit. Hand switches 1 and 2 and the part presence switch must be closed before the drill motor will be turned on. This will assure that there is a part in the machine and that both of the operators hands are in a safe location.

Note that the switches were programmed as normally open contacts. They are all on the same rung (series). All will have to be on for the output to turn on. If there is a part present and the operator puts his/her hands on the start switches, the drill press will run.

If the operator removes one hand to wipe the sweat from his or her brow, the press will stop. Contacts in series such as this can be thought of as logical "AND" conditions. In this case, the part presence switch "AND" the left-hand switch "AND" the right-hand switch would have to be closed to run the drill press.

Study the examples in Figure 3-13 and determine the status of the outputs.

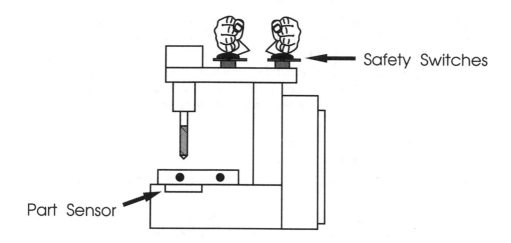

Safety Switches

Part Sensor

Figure 3-12. A simple drilling machine. There are two hand safety switches and one part sensor on the machine. Both hand switches and the part sensor must be true for the drill press to operate. This assures that the operator's hands are not in the way of the drill. This is an "and" condition. Switch 1 and switch 2 and the part sensor must be activated to make the machine operate. The ladder for the PLC is shown in Figure 3-11.

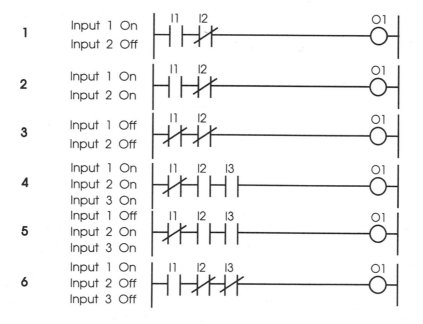

Figure 3-13.

Answers to the ladder logic shown in Figure 3-13.

1. The output for rung 1 will be on. Input 1 is on which closes contact 1. Input 2 is off, so normally closed contact 2 is still closed. Both contacts are closed, so the output is on.

2. The output in rung 2 is off. Input 1 is on, which closes normally open contact 1. Input 2 is on, which forces normally closed contact 2 open. The output cannot be on, because normally closed contact 2 is forced open.

3. The input in rung 3 is on. Inputs 1 and 2 are off so that normally closed contacts 1 and 2 remain closed.

4. The output in rung 4 is off. Input 1 is on which forces normally closed contact 1 open.

5. The output in rung 5 is on. Input 1 is off so normally closed contact 1 remains closed. Inputs 2 and 3 are on, which forces normally open contacts 2 and 3 closed.

6. The output in rung 6 is on. Input 1 is on forcing normally open contact 1 closed. Inputs 2 and 3 are off which leaves normally closed contacts 2 and 3 closed.

Branching

There are often occasions when it is desired to turn on an output for more than one condition. For example, in a house, the doorbell should sound if either the front or rear door button is pushed (the two conditions under which the bell should sound). The ladder would look similar to Figure 3-14. This is called a <u>branch</u>. There are two paths (or conditions) that can turn on the doorbell. (This can also be called a logical "or" condition.)

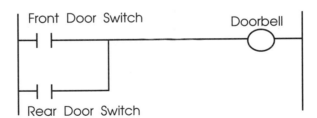

Figure 3-14. This figure shows a parallel condition. If the front door switch is closed the doorbell will sound, "or" if the rear door switch is closed the doorbell will sound. These parallel conditions are also called "or" conditions.

If the front door switch is closed, electricity can flow to the bell. Or if the rear door switch is closed, electricity can flow through the bottom branch to the bell. Branching can be thought of as an "or" situation. One branch "or" another can control the output. "Or's" allow multiple conditions to control an output. This is very important in industrial control of systems. Think of a motor that is used to move the table of a machine. There are usually two switches to control table movement: a jog switch and a feed switch (see Figure 3-15). Both switches are used to turn on the same motor. This is an "or" condition. The jog switch "or" the feed switch can turn on the table feed motor. Evaluate the ladder logic shown in Figure 3-16.

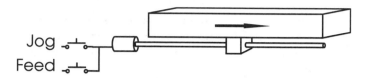

Figure 3-15. Conceptual drawing of a mill table. Note that there are two switches connected to the motor. These represent "OR" conditions. The jog switch or the feed switch can move the table.

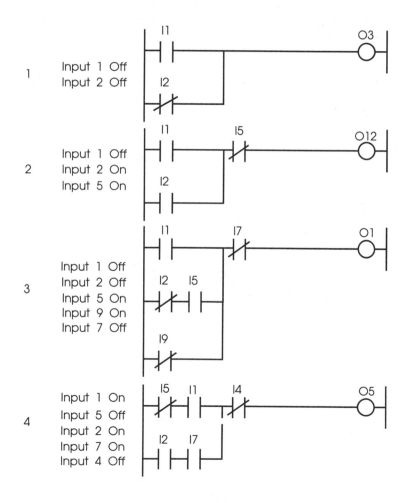

Figure 3-16. The answers are given on the next page.

These are the answers to the examples in Figure 3-16.

1. The output in example 1 would be on. Input 2 is off, so that normally closed contact 2 is closed, allowing the output to be on.

2. The output in example 2 is off. Input 5 is on, which forces normally closed contact 5 open, so the output cannot be on whether or not input 1 "or" input 2 is on. It should also be noted that in these branching examples we have combinations of "ands" and "ors." In English, this example would be: input 1 "and" input 5 "or" input 2 "and" input 5 will turn on output 12.

3. The output in example 3 will be on. Input 2 is off, which leaves normally closed contact input 2 closed "and" input 5 is on, which closes normally open input 5 "and" normally closed input 7 is off, which leaves normally closed contact 7 closed which turns on output 1. In this ladder there are 3 "or" conditions and combinations of "ands."

4. In example 4, the output will be on. Input 5 is off which leaves normally closed contact 5 closed "and" input 1 is on, which forces normally open contact 1 closed "and" input 4 is off, which leaves normally closed contact 4 closed "and" input 4 is off, which leaves normally closed contact 4 closed, turning on the output. Inputs 2 and 7 are also both on, closing normally open contacts 2 "and" 7.

Start/Stop Circuit

Start/stop circuits are extremely common in industry. Machines will have a start button to begin a process and a stop button to shut off the system. Several important concepts can be learned from the simple logic of a start/stop circuit.

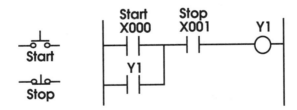

Figure 3-17. Start/stop circuit.

Examine Figure 3-17. Notice that the actual start switch is a normally open pushbutton. When it is depressed, it closes the switch. When the button is released, the switch opens. The stop switch is a normally closed switch. When pressed, it opens.

Now examine the ladder. When the start switch is momentarily depressed, power passes through X000. Power also passes through X001, because the real-world stop switch is a normally closed switch. The output (Y1) is turned on. Note that Y1 is then also used as an input on the second line of logic. Output Y1 is on so contact Y1 also closes. This is called latching. The output latches itself on even if the start switch opens. Output Y1 will shut off only if the normally closed stop switch (X001) is pressed. If X001 opens, then Y1 is turned off. The system will require the start button to be pushed to restart the system. Note that the real-world stop switch is a normally closed switch, but that in the ladder, it is programmed normally open. This is done for safety.

There are as many ways to program start/stop circuits (or ladder diagrams in general) as there are programmers. Safety is the main consideration.

Special Contacts

There are many special purpose contacts available to the programmer. The original PLCs did not have many available. Sharp programmers used normally open and normally closed contacts in ingenious ways to turn outputs on for one scan, to latch outputs on, and so on. PLC manufacturers added special contacts to their ladder programming languages to meet these needs. The programmer can now accomplish these special tasks with one contact instead of a few lines of logic.

Immediate Instructions

Immediate instructions are used when the input or output being controlled is very time dependent. For example, for safety reasons we may have to update the status of a particular input every few milliseconds. If our ladder diagram is 10 milliseconds long, the scan time would be too slow. This could be dangerous. The use of immediate instructions allows inputs to be updated immediately as they are encountered in the ladder. The same is true of output coils.

Transitional Contacts

Transitional contacts are one type of special contact. They are also called one-shot contacts. The symbol for this type of contact is the normal contact symbol plus an arrow pointing either up or down (see Figure 3-18). A down arrow means that when this contact is energized it will transition from high to low for one scan. An arrow pointing up means that when this type of contact is energized it will transition from off to on for one scan. That is why they are called one-shots. They are only active for one scan when energized.

There are many reasons to use this type of contact. They are often used to provide a pulse for timing, counting, or sequencing.

They are also used when it is desired to perform an instruction only once (not every scan). For example, if an add instruction was used to add two numbers once, it would not be necessary to add them every scan. A transitional contact would assure that the instruction executes only on the desired transition.

Figure 3-18. Two transitional contacts: a low-to-high transitioning contact and a high-to-low transitioning contact.

Latching Instructions

Latches are used to lock in a condition. For example, if an input contact is on for only a short time, the output coil would be on for the same short time. If it were desired to keep the output on even if the input goes low, a latch could be used. This can be done by using the output coil to latch itself on (see Figure 3-19).

It can also be done with a special coil called a latching coil (see Figure 3-20). When a latching output is used, it will stay on until it is unlatched. Unlatching is done with a special coil called an unlatching coil. When it is activated the latched coil of the same number is unlatched.

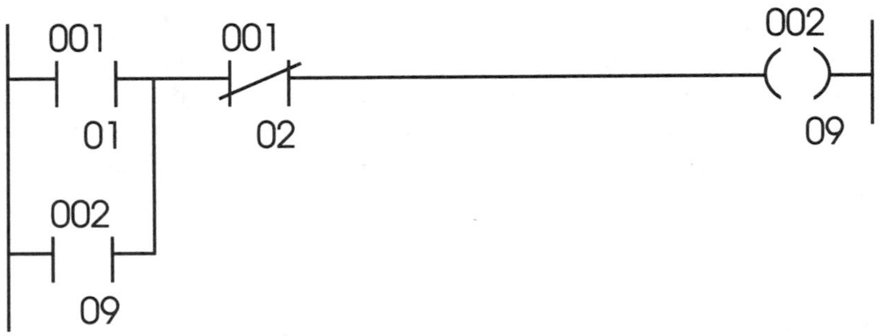

Figure 3-19. This figure shows an example of latching an output on. If input 00101 is true, coil 00209 will be energized. (Remember that contact 00102 is normally closed.) When coil 00209 energizes, it latches itself on by providing a parallel path around 00101. The only way to turn the latched coil off would be to energize normally closed contact 00102. This would open the rung and deenergize coil 00209.

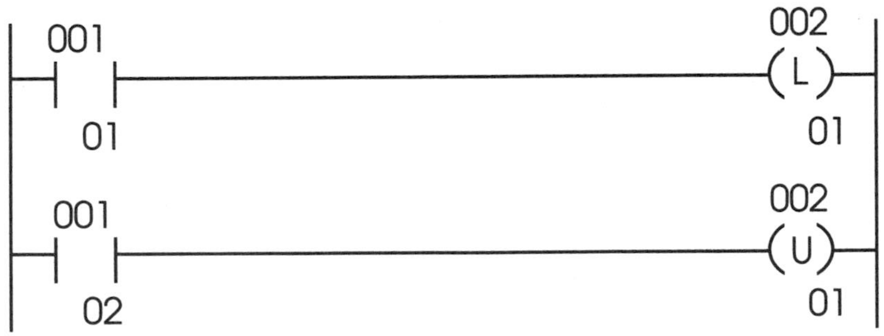

Figure 3-20. Use of a latching output. If input 00101 is true, output 00201 energizes. It will stay energized even if input 00101 becomes false. 00201 will remain energized until input 00102 becomes true and energizes the unlatch instruction (coil 00201). Note that the coil number of the latch is the same as that of the unlatch. This is an example of an Allen Bradley latch instruction.

Chapter 3: Fundamentals of Programming

Master Control Relays

Master control relays are used to control blocks of a ladder diagram. They can control entire sections of a program or the whole program. The master control relay should not be confused with the hardwired master control relay that should be used to protect every application. Hard-wired master control relays are used to provide for the immediate shutdown of power in the event of an emergency condition.

The master control relay instruction can be used to control one section or all of a ladder diagram (see Figure 3-21). The use of a master control relay normally requires the use of a MCR instruction to begin the controlled area and another MCR instruction to show the end of the area. The beginning MCR rung determines whether or not the MCR logic is active. If the inputs on the rung of the initial MCR instruction are true, the MCR is active. Each rung controls its outputs, just as in normal operation. If the initial MCR instruction rung is false, all nonretentive output instructions are deenergized regardless of their rung condition.

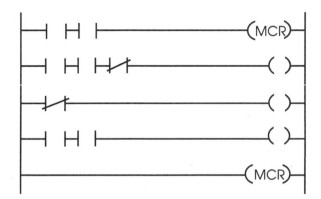

Figure 3-21. Use of a master control relay (MCR). Note that there are no input conditions for the second MCR. The second MCR marks the end of the controlled zone.

Zone Control Logic Instructions

Allen Bradley also has zone control logic instructions (ZCL) available. The programming is very similar to the MCR. There is a ZCL to begin the controlled section of the program and one to end it. The difference is in the way the outputs react. When the rung conditions for the ZCL are true, the ZCL is active. While the zone is active the actual rung conditions in the zone control their outputs (see Figure 3-22).

In zone control logic, if the ZCL input conditions go false the outputs within the zone retain the state they were in. These outputs may now be controlled by other sections of ladder or other zones. Zone control logic can be used to break programs into sections of logic. Remember that most of the time used in writing a program is devoted to interlocking the logic. (Making sure that outputs are not on when they should not be.) The use of zones can simplify this process by reducing the task of interlocking. Output logic is active only in the zones where the ZCL input conditions have been met.

Figure 3-22. Use of zone control logic (ZCL). Note that the ZCL that marks the end of the area is unconditional.

PLC Scanning and Scan Time

Now that you are familiar with some basic PLC instructions and programming, it is important to understand the way a PLC executes a ladder diagram. Most people would like to believe that a ladder is a very sequential thing. We like to think of a ladder as first things first. We would like to believe that the first rung is evaluated and acted on before the next, and so on. We would like to believe that the CPU looks at the first rung, goes out and checks the actual inputs for their present state, comes back, immediately turns on or off the actual output for that rung, and then evaluates the next rung. This is not exactly true. Misunderstanding the way the PLC scans a ladder can cause programming bugs.

Scan time can be divided into two components: I/O scan and program scan. When the PLC enters run mode it first takes care of the I/O scan (see Figure 3-23). The I/O scan can be divided into the output step and the input step. During these two steps the CPU transfers data from the output image table to the output modules (output step) and then from the input modules to the input image table (input step).

The third step is logic evaluation. The CPU uses the conditions from the image table to evaluate the ladder logic. If a rung is true, the CPU writes a one into the corresponding bit in the output image table. If the rung is false, the CPU writes a zero into the corresponding bit in the output image table. Note that nothing concerning real-world I/O is occurring during the evaluation phase. This is often a point of confusion. The CPU is basing its decisions on the states of the inputs as they existed before it entered the evaluation phase. We would like to believe that if an input condition changes while the CPU is in the evaluation phase, it would use the new state. It cannot. (Note: It actually can if special instructions called immediate update contacts and coils are used. For the most part most ladders will not utilize immediate instructions. This is covered in more detail later.) The states of all inputs were frozen before it entered the evaluation phase. The CPU does not turn on/off outputs during this phase either. This phase is only for evaluation and updating the output image table status.

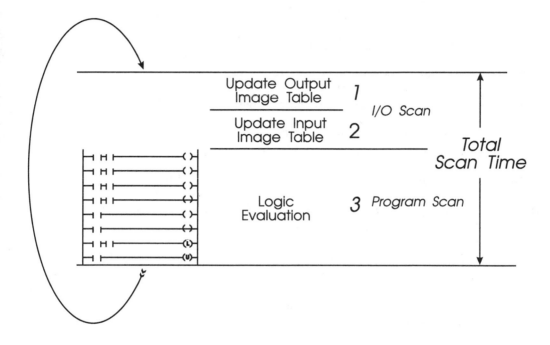

Figure 3-23. A generic example of PLC scanning.

Once the CPU has evaluated the entire ladder, it performs the I/O scan again. During the I/O scan the output states of real-world outputs are changed depending on the output image table. The real-world input states are then transferred again to the input image table.

All of this only took a few milliseconds (or less). PLCs are very fast. That is why troubleshooting can be so troublesome. Scan time is the sum of the times it takes to execute all of the individual instructions in the ladder. Simple contacts and coils take very little time. Complex math statements and other types of instructions take much more time. Even a long ladder diagram will normally execute in less than 50 milliseconds or so. There are considerable differences in the speeds of different brands and models of PLCs. Manufacturers normally will give scan time in terms of fractions of milliseconds per "K" of memory. This can help give a rough idea of the scan times of various brands.

Questions

1. What is a contact? A coil?

2. What is a transitional contact?

3. What are transitional contacts used for?

4. Explain the term <u>normally open</u> (<u>examine on</u>).

5. Explain the term <u>normally closed</u> (<u>examine off</u>).

6. What are some uses of normally open contacts?

7. Explain the terms <u>true</u> and <u>false</u> as they apply to contacts in ladder logic.

8. Design a ladder that shows series input (AND logic). Use X5, X6, AND NOT (normally closed contact) X9 for the inputs and use Y10 for the output.

9. Design a ladder that has parallel input ("OR" logic). Use X2 and X7 for the contacts.

10. Design a ladder that has three inputs and one output. The input logic should be: X1 AND NOT X2, OR X3. Use X1, X2, and X3 for the input numbers and Y1 for the output.

11. Design a three-input ladder that uses "and" logic and "or" logic. The input logic should be X1 OR X3, AND NOT X2. Use contacts X1, X2, and X3. Use Y12 for the output coil.

12. Design a ladder in which coil Y5 will latch itself in. The input contact should be X1. The unlatch contact should be X2.

13. Design a latching circuit using an Allen Bradley latch and unlatch instruction. Use contact 00107 for the latch input, 00102 for the unlatch input, and 00513 for the coil.

14. What is the primary purpose of MCRs?

15. Is a ladder logic MCR sufficient to guarantee the safety of a system? Why or why not?

16. Explain the primary purpose of zone control logic.

17. Draw a diagram and thoroughly explain what occurs during a PLC scan.

18. Examine the rungs below and determine whether the output for each is on or off. The input conditions shown represent the states of real-world inputs.

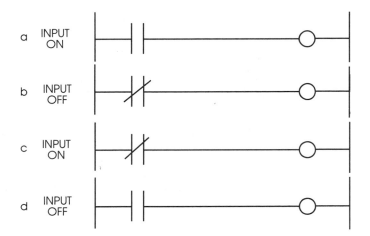

19. Examine the rungs below and determine whether the output for each is on or off. The input conditions shown represent the states of real-world inputs.

20. Examine the rungs below and determine whether the output for each is on or off. The input conditions shown represent the states of real-world inputs.

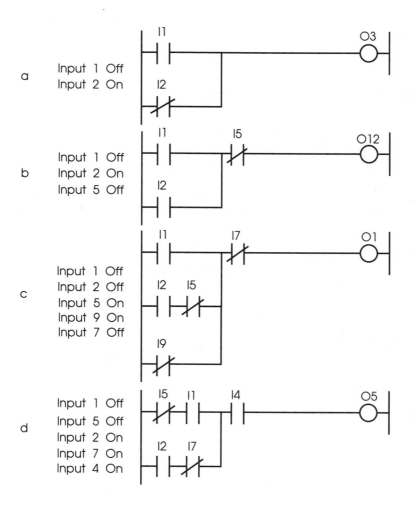

Additional Exercises

1. Study the TISOFT programming software tutorial. It is available from your instructor.
   a. What two values must be entered for a timer?
   b. What does the scan cycle for the 405 series look like?
   c. Describe how immediate contacts work.
   d. What types of information can be printed.

Chapter 4

Timers and Counters

Timers and counters are invaluable in PLC programming. Industry has a need to count product, time sequences, and so on. In this chapter we will examine the types and programming of timers and counters. Timers and counters are similar in all PLCs. In this chapter we will show examples of several of the leading brands.

Objectives

Upon completion of this chapter, the student will be able to:

Describe the use of timers and counters in ladder logic .

Define such terms as <u>retentive</u>, <u>cascade</u>, <u>delay-on</u>, <u>delay-off</u>, <u>flow diagram</u>, and <u>pseudocode</u>.

Develop flow diagrams and pseudocode for applications.

Utilize timers and counters to develop applications.

Timers

Timing functions are very important in PLC applications. Cycle times are critical in many processes. Timers are used to delay actions. They may be used to keep an output on for a specified time after an input turns off or to keep an output off for a specified time before it turns on.

Think of a garage light. It would be a nice feature if a person could touch the on switch and the light would immediately turn on, and then stay on for a given time (maybe 2 minutes). At the end of the time, the light would turn off. This would allow you to get into the house with the light on. In this example, the output (light) turned on instantly when the input (switch) turned on. The timer counted down the time (timed out) and turned the output (light) off. This is an example of a <u>delay-off</u> <u>timer</u>.

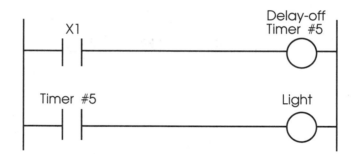

Figure 4-1. Delay-off timing circuit. If contact X1 closes, delay-off timer immediately turns on which turns on the light. When the timer reaches the programmed time it will turn off, which turns the light off also.

Consider Figure 4-1. When switch X1 is activated, the timer turns on and starts counting. The timer is used in the second rung as a contact. If the timer is on, the contact in rung 2 (timer 5) closes and turns on the light. When the timer times out (in this case 2 minutes), the contact in rung 2 opens and the light turns off. As you see from this example, output coils can be used as contacts to control other outputs.

This type of timer is called a <u>delay-off timer</u>. The timer turns on instantly, counts down and then turns off (delay-off).

The other type of timer is called <u>delay-on</u>. When input X3 to the delay-on timer is activated, the timer starts counting, but remains off until the time has elapsed (see Figure 4-2). In this case, the switch is activated and the timer starts to count, but it remains off until the total time has elapsed. Then the timer turns on, which closes the contact in rung 2 and turns on the light.

Many PLCs use block-style timers and counters (see Figure 4-3). No matter what brand of PLC it is, there are many similarities in the way timers are programmed. Each timer will have a number to identify it. For some it will be as simple as T7 (timer 7). For others it may be the address of the storage register that holds the accumulated value of the timer, such as storage register 4001.

Each timer will have a time base. Timers can typically be programmed with several different time bases: 1 second, 0.1 second and 0.01 second are typical time bases. If a programmer entered .1 for the time base and 50 for the number of delay increments the timer would have a 5-second delay (50 X 0.1 second = 5 seconds).

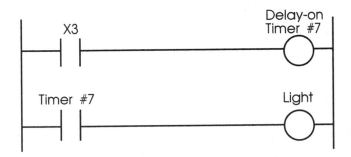

Figure 4-2. Delay-on timing circuit. In this example if contact X3 closes the timer will begin timing. When the time reaches the programmed time the timer will turn on which will turn the light on.

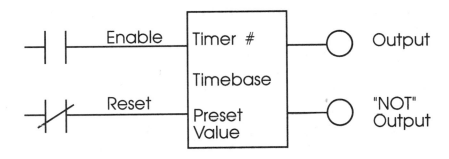

Figure 4-3. This figure shows a typical block type timer.

Timers also must have a preset value. The preset value is the number of time increments the timer must count before changing the state of the output. The actual time delay would equal the preset value multiplied by the timebase. Presets can be a constant value or a variable. If a variable is used, the timer would use the real-time value of the variable to calculate the delay. This allows delays to be changed depending on conditions during operation. An example is a system that produced two different products, each requiring a different time in the actual process. Product A requires a 10-second process time, so the ladder logic would assign ten to the variable. When product B comes along the ladder logic can change the value to that required by B. When a variable time is required, a variable number is entered into the timer block. Ladder logic can then be used to assign values to the variable.

Timers typically have two inputs. The first is the timer enable input. When this input is true (high) the timer will begin timing. The second input is the reset input. This input must be in the correct state for the timer to be active. Some brands of PLCs require that this input be low for the timer to be active: other manufacturers require a high state. They all function in essentially the same manner, however. If the reset line changes state, the timer clears the accumulated value. For example, the timer in Figure 4-3 requires a high for the timer to be active. If the reset line goes low, the timer clears the accumulated time to zero.

Timers can be retentive or nonretentive. <u>Retentive timers</u> do not lose the accumulated time when the enable input line goes low. They retain the accumulated time until the line goes high again. They then add to the count when the input goes high again. Nonretentive timers lose the accumulated time every time the enable input goes low. If the enable input to the timer goes low, the timer count goes to zero.

Allen Bradley PLC-2 Timers

Allen Bradley timers require two values: an accumulated value (AC) and a preset value (PR). Each value requires one word of memory. The accumulated value holds the current number of elapsed time intervals. The upper four bits (most significant bits) are used as status bits. The preset value is the number of time intervals it is desired to time to. When the number of preset intervals equal the accumulated value of time intervals, a status bit is set. The status bit can turn devices on or off. Bit number 15 is the timed bit. Bit 15 is set to either a one or a zero, depending on which timer instruction is chosen when the timer has timed out. Bit number 17 is the enable bit. It is set to one when the rung conditions are true and reset to zero when the rung conditions are false. There are three time intervals that can be selected: 1.0 second, 0.1 second, or 0.01 second.

There are four types of timer instructions available: on-delay -(TON)-, off-delay -(TOF)-, retentive -(RTO)-, and retentive timer reset -(RTR)-.

On-Delay Timer -(TON)-

The Allen Bradley on-delay timer (TON) is programmed like an output instruction. The TON timer does not accumulate the time if the timer loses its input signal. It will continue to count time intervals as long as the input condition is true. The timer will be "on" when the accumulated time intervals are equal or greater than the preset. As soon as the input to the timer goes low the accumulated value is set to zero and the timer is "off."

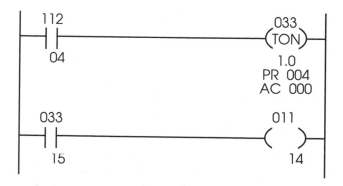

Figure 4-4. Use of an Allen Bradley TON timer in a ladder diagram.

Figure 4-4 shows the use of a TON timer in a ladder diagram. The timer number is 33. If input 11204 becomes true, the timer will start to accumulate time.

Note that a 1.0-second time base is used. This means that the actual time delay will be the preset (4) multiplied by the time base (1.0 second). The delay will be 4 seconds. When the accumulated time equals the preset time, bit number 15 (03315 - timer 33 bit 15) will be set to one. The timer is then used as an input on the second rung to control output coil 01114. This means that when the timer times out, output 01114 will be energized.

Off-Delay Timer -(TOF)-

The delay-off timer (TOF) is programmed like an output instruction also. The main difference is that the timer increments the accumulated time when the input condition is false. The Allen Bradley TOF timer does not accumulate the accumulated time if the input condition changes state. The accumulated value of the timer resets to zero if the input condition becomes true. The timer counts time intervals as long as the input condition remains false. In this manner the output will turn on instantly when the input condition becomes true and will remain on until the accumulated time equals the preset time.

Bit numbers 15 and 17 will be set as soon as the rung conditions become true. The timer will then begin to increment the accumulated (AC) value. When the accumulated value equals the preset value, bit number 15 will be reset. Bit number 17 is reset when the rungs input condition becomes false.

Retentive Delay-On Timer -(RTO)-

The retentive delay-on timer (RTO) is very similar to the regular delay-on instruction. The only difference is that the accumulated time does not get reset if the input condition becomes false. The accumulated value remains at the present value until the input condition becomes true again. The timer then begins to add time interval counts to the accumulated value of the timer.

When the rung conditions become true, the timer will begin to count time base intervals. Bit number 15 will be set when the accumulated value equals the preset value (ACC = PRE), and the timer stops timing. Bit number 17 is set when the rung becomes true. When the rung's input condition becomes false, the accumulated value is retained. Bit number 15's status will not be changed. Bit number 17 will be reset.

The timer accumulated value is reset to zero only if a special timer instruction is used. The instruction is called a retentive timer reset instruction.

Retentive Timer Reset -(RTR)-

The retentive timer reset (RTR) must be given the same number as the timer that it is supposed to reset. If the input conditions to the RTR become true, the instruction will reset the accumulated value of its associated timer to zero. It will also reset the status bits for the associated timer. This means that bit numbers 15 and 17 will be reset to zero. The retentive timer reset is the only way to reset the accumulated value of a retentive timer instruction.

See Figure 4-5 for examples of timer use.

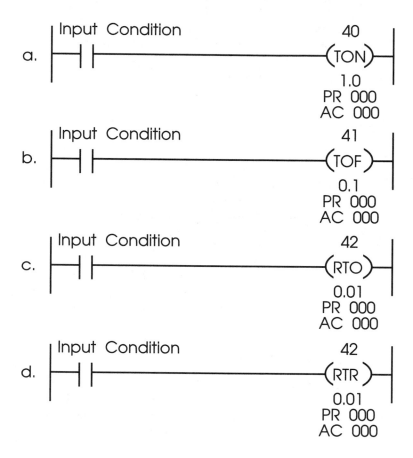

Figure 4-5. Use of Allen Bradley timer instructions: (a) shows the use of a TON instruction; (b) shows the use of a TOF instruction; (c) shows the use of a RTO instruction; (d) shows the use of a RTR instruction.

Gould Modicon Timers

Figure 4-6 shows a typical Gould Modicon timer block with a typical enable input and reset input. The preset value for the timer must be entered. The time base must be chosen. The actual delay can be calculated by multiplying the preset value by the time base. The timer number is actually a storage register where the accumulated value of the timer will be stored. There are two outputs possible from the timer. The first output is one that will turn on when the timer reaches the timer preset value. The second output is the opposite. It is "normally on" and it will turn off when the timer reaches the preset value. These two outputs make it very easy to program delay-on or delay-off logic.

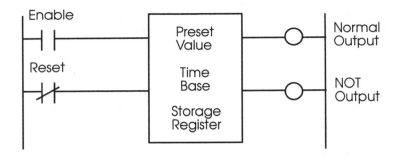

Figure 4-6. Gould Modicon timer block.

Figure 4-7 shows the actual use of a Gould Modicon timer. Note the first number of the inputs, outputs, and storage register. The 1 in the inputs denotes that the number will be an input. The first 0 in output 0005 shows that the number represents an output. The 4 of the number 4005 denotes shows that the number is a storage register. The time base is 0.1 second. The preset is 250, so the value of the delay is 25 seconds (250 X 0.1 second = 25 seconds). It will be used as a delay-on timer because the top output on the timer block was used. Note that input 7 is programmed as a normally closed contact. The reset line must be high for the timer to time. If this line goes low the timer is reset to a count of zero and is unable to time again until the reset line goes high. Input 2 is being used as the enable line. The current value of the timer will be stored in register 4004. If the timer value reaches 25 seconds, output 5 will turn on and stay on. If input 2 goes low or input 7 goes low, output 5 will turn off.

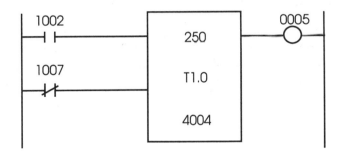

Figure 4-7. Gould Modicon timer example.

Omron Timers

There are two types of timers: TIM and TIMH. The TIM timer measures in increments of 0.1 second. It is capable of timing from 0 to 999.9 seconds with an accuracy of plus or minus 0.1 second. The high-speed timer (TIMH) measures in increments of 0.01 second. Both timers are decrementing-style delay on timers. They require a timer number and a set value (SV). When the set value (SV) has elapsed, the timer output turns on. Timer counter numbers refer to an actual address in memory. Numbers must not be duplicated. You cannot use the same number for a timer and a counter. Timers/counters are numbered 0 to 511.

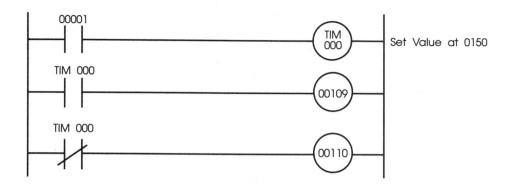

Figure 4-8. Omron TIM timer being used as a normally open contact and a normally closed contact.

Figure 4-8 shows the use of a TIM timer in a ladder diagram. When the timer decrements to zero the timer output turns on. Output 00110 will be on until the timer becomes true. Note that a normally closed timer 000 has been used as the contact for output coil 00110. This means that as long as the timer is false, output 001100 will be energized. When the timer becomes true output 00109 will turn on. This, of course, also means that output 00110 will deenergize. The timer will reset if input 00001 becomes false. A power failure would also cause the timer to reset.

This can be solved if a retentive timer is used. A counter is used to make a retentive timer. Remember that there is very little difference between a timer and a counter. Both count increments or events. If a counter is used to count time increments, it becomes a timer.

Figure 4-9 shows the use of a counter to make a retentive timer. A special bit is being used to create the 1-second pulse. Bit 25502 is a special bit that Omron provides. It is a one second clock pulse. If input 00001 is present and the one second pulse bit makes 20 transitions, the counter will be true. Even if power is lost the present value of the count is retained. This means that we now have a retentive timer.

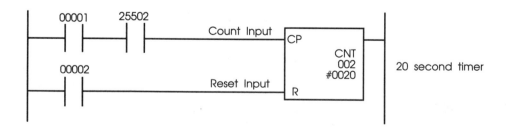

Figure 4-9. This figure shows the use of a counter to create a retentive timer. This will preserve the present value of the timer in the event of a power interruption. Note that contact 25502 is a special internal bit. Bit 25502 is a 1-second timing pulse that can be used by the programmer.

The high-speed timer (TIMH) times in 0.01-second increments. The TIMH timer has a range of 0.00 to 99.99 seconds. Scan time can affect TIMH timers numbered 48 to 511. Timers 48 to 511 may be inaccurate if the scan time exceeds 10 milliseconds. If the scan time is more than 10 milliseconds use timer numbers below 48.

Figure 4-10 shows the use of a TIMH timer. It operates just like the TIM timer except that the time increment is smaller. Remember that the TIM and TIMH timers are not retentive, if the enable input is lost the present value is lost.

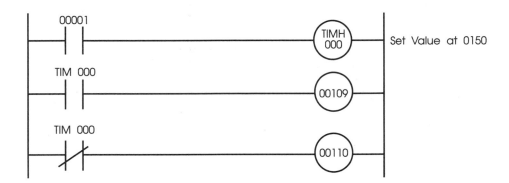

Figure 4-10. Omron high-speed timer (TIMH). Output 00110 will be on until the timer times out. This is the case because a normally closed contact was used. When the timer times out, output 00109 will be energized.

Square D Timers

Figure 4-11 shows a typical Square D timer. There are two inputs: a timing line (enable) and a clear (reset) line. There is a storage register (sometimes called data register) address. The storage register address is where the current value of the timer will be stored. There is a time base (0.1 second, 0.01 second, and 0.1 minute). There is a preset value which is multiplied by the time base to calculate the time delay. When the number in the storage register equals the preset value the output is turned on. In this case when storage register 17 equals 50, the output will turn on. Square D calls the preset a decode value. There is also an output shown at the bottom of the block. This is the output number that will turn on when the timer reaches the required delay.

Figure 4-12 shows a Square D retentive timer used as a nonretentive timer. This same method can be used with most timers. The enable input to the timer has been programmed without a contact. This means that this input is always high. The counter is always enabled. This means that input 01-05 totally controls the timer. If input 01-05 is off, register S17 is reset to zero. If 01-05 is high, the timer begins to time. If register S17 equals 50 then output 04-10 will turn on. It will stay on until input 01-05 goes low. If input 01-05 goes low before S17 reaches 50 the output will stay off. As you can see, even though it is a retentive timer, it never retains the count if the input (01-05) goes low.

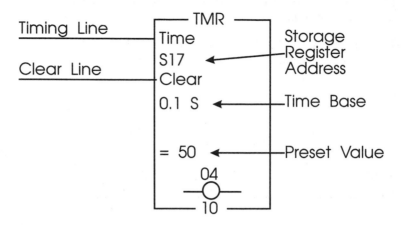

Figure 4-11. Square D timer.

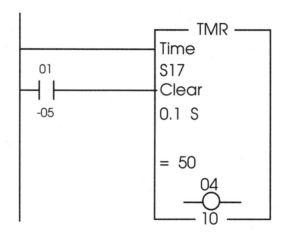

Figure 4-12. Square D timer. This timer is being used as a delay-on
timer. When input 01-05 is true (on), the timer counts. If the timer reaches 5
seconds (50 X 0.1) output 04-10 will be turned on until input 01-05 turns off.

Siemens Industrial Automation Timers

Figure 4-13 shows a Siemens Industrial Automation, Inc. nonretentive timer. There are two
values required for the timer block. The first value is the timer number. The second is the preset
value. If X0 becomes true (high), the timer will begin to time. If the time equals the preset, the

timer contact turns on. The timer contact can then be used in another rung. In this case if the timer reaches its preset, output Y1 will turn on until X0 goes low again.

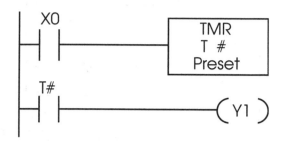

Figure 4-13. Siemens Industrial Automation series 405 timer.

Figure 4-14 shows a Siemens Industrial Automation accumulating timer. Accumulating timers can time up to a maximum of 99,999,999. Note that the TMRA in the block means "accumulating timer." The Siemens Industrial Automation, Inc. series 405 timer requires a low at the reset line to activate the timer. If the reset line goes high, the timer resets. In this case if X0 goes high, the timer will begin to time. If the timer count reaches the preset value, the timer contact will turn on. This would turn on output Y1. If X0 goes low before the count reaches the preset time, the timer will hold the present time and will not reset the time count to zero. When X0 goes high again the timing will continue. An accumulating timer will never reset to zero unless the reset line goes high.

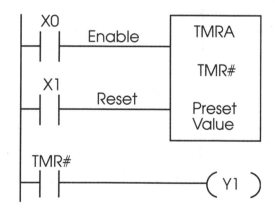

Figure 4-14. Siemens Industrial Automation accumulating timer example.

Figure 4-15 shows the use of a Siemens Industrial Automation timer. Note that this is a fast accumulating timer (TMRAF). This is really a retentive timer with a time base of 0.01 second. The timer number is 5. The enable input is X0. The reset input is X1. T5 controls output Y1. (If the timer preset has been reached, the timer is on and output Y1 is on.) The preset in this case is a variable (V1400). The value for the preset value will equal whatever the value of V1400 is. Other ladder logic would establish the value of V1400.

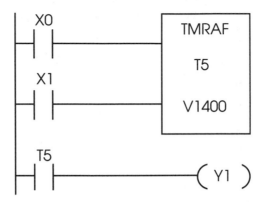

Figure 4-15. This figure shows a Siemens Industrial Automation fast accumulating timer (TMRAF). The time increment for the fast timer is 0.01 second.

The accumulated value of timers is kept in variables. There are 128 (0 to 177 octal) timers available. Their accumulated value is stored in variables V00000 to V00177. If we wanted to monitor the accumulated value of timer 5 we would examine variable V00005. These accumulated values can be used in ladder logic just as any other variable could be.

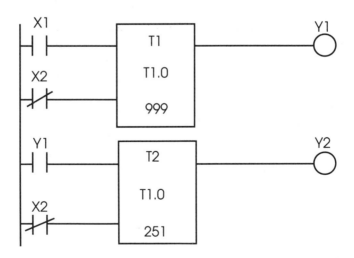

Figure 4-16. Two timers used to extend the time delay. The first timer output, Y1, acts as the input to start the second timer. When input X1 becomes true, timer 1 begins to count to 999 seconds. (The limit for these timers is 999 seconds. The limit for accumulating timers is 99,999,999.) When it reaches 999, output Y1 turns on. This activates input Y1 to timer 2. Timer 2 then counts to 251 seconds and then turns on output Y2. The delay was 1250 seconds.

Chapter 4: Timers and Counters

Cascading Timers

Applications sometimes require longer time delays than one timer can accomplish. Multiple timers can than be used to achieve a longer delay than would otherwise be possible. One timer acts as the input to another. When the first timer times out it becomes the input to start the second timer timing. This is called <u>cascading</u> (see Figure 4-16).

Counters

Counting is very important in industrial applications. Often, the product must be counted so that another action will take place. For example, if 24 cans go into a case, the twenty-fourth can should be sensed by the PLC and the case should be sealed. Counters are required in almost all applications.

Several types of counters are available, including up counters, down counters, and up/down counters. The choice of which to use is dependent on the task to be done. For example, if we are counting the finished product leaving a machine, we might use an up counter. If we are tracking how many parts are left, we might use a down counter. If we are using a PLC to monitor an automated storage system, we might use an up/down counter to track how many are coming and how many are leaving to establish an actual, total number in stock.

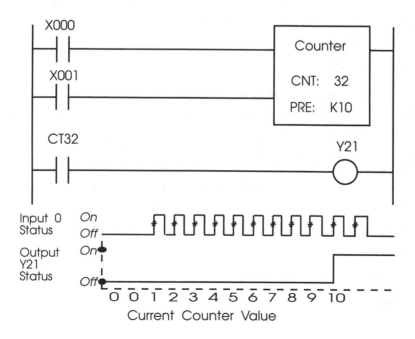

Figure 4-17. How a typical counter works. When 10 or more low-to-high transitions of input X000 have been made, counter CT32 is energized, which energizes output Y21.

Counters normally use a low-to-high transition from an input to trigger the counting action (see Figure 4-17). They also have a reset line to clear the accumulated count. There are actually a lot of similarities to timers. Timers count the number of time increments and counters count the number of low-to-high transitions on the input line.

Note that this timer is edge-sensitive triggered. The rising, or leading edge, triggers the counter. Other devices may be level (magnitude) sensitive. X000 is used to count the pulses. Every time there is an off-to-on transition on X000, the counter adds one to its count. When the count equals the preset value, the counter turns on, which turns on output Y20. X001 is used as a reset/enable. If contact X001 is closed, the counter is returned to zero. The counter is active (enabled or ready to count) only if X000 is off (open).The example just described is an up counter.

Down counters cause a count to decrease by one every time there is a pulse. There are also up/down counters. An up/down counter has one input that causes it to increment the count and another that causes it to decrement the count.

There are ladder diagram statements that can utilize these counts for comparing and/or decision making. Counts can also be compared to constants or variables to control outputs.

Allen Bradley PLC-2 Counters

The Allen Bradley counter is used to count events. (Note: PLC-5 and SLC PLC instructions are covered in Appendix A. In this chapter we focus on PLC-2s.) Events are low-to-high transitions of the input conditions to the counter. The counter instructions are very similar to the AB timer instructions. They are programmed like output instructions. The upper 4 bits in the accumulated value (AC) are status bits.

Bit number 14 is the overflow/underflow bit. It is set to a 1 when the accumulated value (ACC) of a CTU counter exceeds 999 or when the ACC value of a CTD instruction falls below zero.

Bit number 15 is the count complete bit. This bit is set to a 1 when the accumulated value is greater than or equal to the preset value (PRE).

Bit number 16 is the enable bit for the CTD instruction. It is set to a 1 when the rung's input condition is true.

Bit number 17 is the enable bit for the CTU instruction. It is set to a 1 when the counter's rung conditions are true.

There are three types of counter instructions:

> *Up-counter*
>
> *Down-counter*
>
> *Counter reset*

Their use is shown in Figure 4-18.

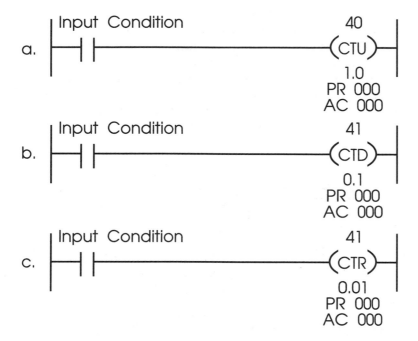

*Figure 4-18. Use of three types of counters. Figure (a) count-up (CTU)
counter. If the input condition makes a transition from low to high, the counter
increments. (b) Count-down counter (CTD). When there is a high-to-low
transition, the counter decrements one from the accumulated count. (c)
Counter reset. If the input condition becomes true, the reset will set the
accumulated count of counter 41 to zero.*

Up-Counter Instruction -(CTU)-

The up-counter instruction (CTU) increments its count for every low-to-high transition of the
counter input condition. There are two values for the instruction: accumulated value and preset
value. The accumulated value is the present count. The preset value is the desired count. When
the counter accumulated value (AC) is equal to or greater than the preset value (PR), the counter
output status bit is set to 1.

Down-Counter Instruction -(CTD)-

The down counter instruction (CTD) is similar but opposite. The counter decrements its value
each time there is a low-to-high transition of the counter input conditions.

Counter Reset Instruction -(CTR)-

The counter reset instruction is used to reset the up-or down-counter accumulated value and
status bits to zero.

Gould Modicon Counters

Figure 4-19 shows an example of a Gould Modicon counter. The counter has two inputs, a enable and a reset. The reset line must be high for the counter to count. Every time the enable input makes a transition from low to high, the counter will increment. The preset value for this counter was set to 6. When the count equals 6, the outputs will change state. The current count is kept in data register 4005. The top output is off unless the actual count is equal to or greater than the preset. The bottom "NOT" output is on unless the count in 4005 is equal or greater than the value of the preset.

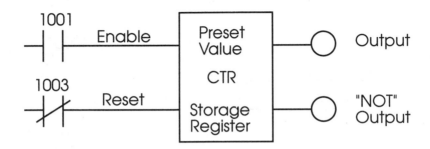

Figure 4-19. Gould Modicon counter. Note that two outputs are available.

Figure 4-20 shows the use of a Gould Modicon counter. Every time input 1001 makes a low-to-high transition the counter will increment the count in data register 4005. When the count reaches 6 or more, output 0005 will turn on. Output 0005 will stay on until the counter is reset by the reset line going low.

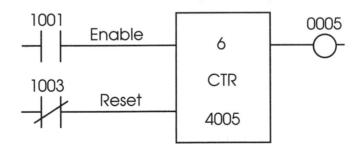

Figure 4-20. Gould Modicon Counter.

Omron Counters

Two types of Omron counters are available: CNT and CNTR. CNT is a decrementing counter and CNTR is a reversible counter instruction. Timers/counters can be numbered from 0 to 511. Numbers cannot be duplicated. If you use number 1 for a counter you may not use it for a timer. Both types of counters require a counter number, a set value (SV), inputs, and a reset input.

Figure 4-21 shows the use of a Omron counter (CNT). The set value is 20. For every low-to-high transition of the counter enable input (00001) the counter decrements the set value. When the count reaches zero the counter output turns on. A low-to-high transition of the reset input will reset the count to the set value.

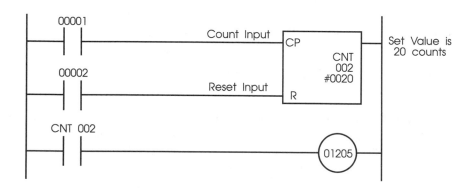

Figure 4-21. Use of a Omron CNT counter. Note that output 01205 will turn on when the counter becomes true. The output will remain on until the reset input to the counter goes high and resets the set value to 20.

There is also a reversible counter type available (CNTR). This counter has three inputs. One input is used to make the counter count up. The second input is used to decrement the count. The third is used to reset the counter. When the reset input is high the present value is set to 0000. Figure 4-22 shows the use of a CNTR counter.

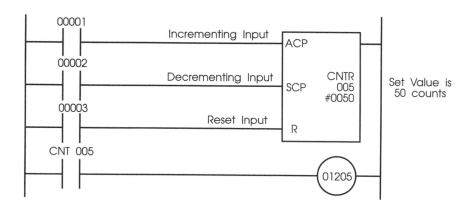

Figure 4-22. Use of a CNTR counter. The counter output will turn on whenever the present count becomes 0000.

Omron timers and counters can use external channels to receive their set values. Instead of giving a set value the programmer assigns a channel from which the timer/counter will "look up" its set value. In this way a timer/counter set value can be made a variable.

Square D Counters

Figures 4-23 and 4-24 illustrate Square D up/down counters. This type of counter has three inputs. The first is the up-count input. A low-to-high transition on this line will increment the count by one (provided the counter is active).

The second input is the down-count input. A low-to-high transition causes the counter to decrement the count by one.

The third input is the clear input. This line must be held high to activate the counter. If the line goes low, the count is cleared.

A storage register address is required. This is where the present value of the count is stored. A preset is required also. The preset is the desired count. In this case when the counter reaches a count of 75, output 03-04 will turn on and remain on until the clear line goes low.

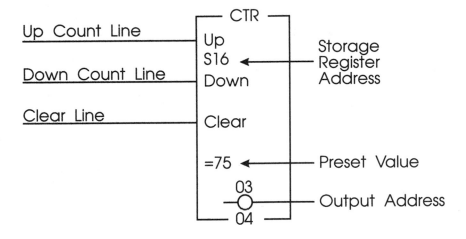

Figure 4-23. Square D counter. The programmer must enter a storage register address, a preset value, and the output address that will be controlled by the counter. The storage address will contain the accumulated count. A low-to-high transition of the up-count line will cause the accumulated value to increase by one. A low-to-high transition of the down count line will cause the accumulated value to decrease by one. Any time the clear line goes low, the accumulated value will be cleared.

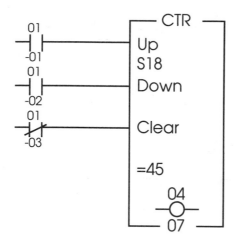

Figure 4-24. Square D counter. Every time input 01-01 makes a low-to-high transition, the accumulated count will increase by one. Any time input 01-02 makes a low-to-high count, the accumulated count is decremented by one. When the accumulated count is equal to or greater than the preset value (45 in this case), output 04-07 will be energized. The accumulated count in storage register S18 can be monitored or used in other instructions. If input 01-03 becomes false, the accumulated value will be cleared.

Siemens Industrial Automation Counters

Siemens Industrial Automation, Inc. counters are very similar to the other brands of counters with one exception (see Figures 4-25 and 4-26). The 405 series uses a normally open contact for the reset line. If the reset line goes high, the counter is reset. The programmer supplies a counter number and a preset value for each counter. The counter number can then be used as a contact. When the counter accumulated count is equal or greater than the preset value, the counter is true. The counter can be reset by making the reset input go high.

The preset value can be a constant (K) or a variable. If it is a variable, the CPU will get the preset value from that variable. If a constant is used, the letter K precedes it. K5 would be a preset value (constant) of 5.

The memory locations that hold the accumulated values can be monitored or used in comparison instructions. The accumulated values are kept in variable memory. For counters the variable numbers are V01000-V01177. Note that there are 128 counters available, or 0 to 177 octal. Note also that there are 128 (V01000 to V01177 octal) counter variables available.

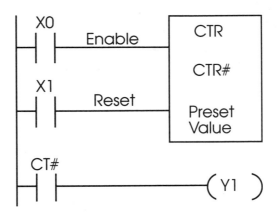

Figure 4-25. Siemens Industrial Automation, Inc. series 405 counter. The programmer must provide a counter number and a preset value. Note that the reset input is a normally open contact. If the reset line becomes high, the counter accumulated value will be set to zero.

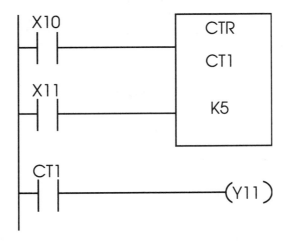

Figure 4-26. Siemens Industrial Automation, Inc. series 405 counter. For every low-to-high transition of input X10, CT1 will increment the count. When the accumulated count is equal to or greater than the preset value (5), counter CT1 will be true. The counter is used as contact CT1. When counter CT1 becomes true, it will energize output Y11.

Programming Hints

Timer and counters are indispensable to application development. There is much in common among PLC manufacturers' implementation of timers and counters. Once the technician becomes familiar with one, the rest become very easy to learn.

When you are asked to program a new brand or type of PLC there are some logical steps that can make the task less frustrating. First make sure that the programming device is really communicating with the PLC. There are several ways to be sure. Most programming software will check and warn you if communications are not established. Some PLC communications modules also have LEDs that flash when communications are taking place. If there is a problem, the most likely place to look is at the cable. Is it the right cable, and is it properly attached to the correct port? The other possible problem is that the PLC may have been assigned a station address different from the one that the software is trying to access. The other potential problem is that the communication parameters were setup incorrectly. Check the baud rates, number of data bits, stop bits, and parity of the software and PLC.

Assuming that the PLC is now communicating, the next step is to see if a simple ladder diagram can be entered and executed. Keep it simple. One contact and one coil are adequate. Make the contact a normally closed contact. Execute the ladder. If the output LED turns on, you have accomplished several important tasks. You have entered a ladder and executed it successfully. You have also figured out the correct I/O numbering.

The next step is probably to get a timer to work, followed by a counter, and so on. Then you can develop the real application quite quickly.

When writing any program there are some useful techniques that make the task less painful. One of these utilizes a <u>flow diagram</u>, which is a pictorial representation of a system or its logic. Figure 4-27 shows a few of the more common flow diagram symbols.

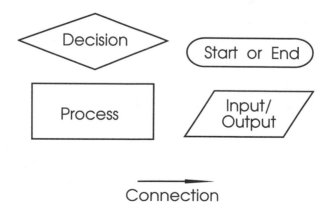

Figure 4-27. Common flow diagram symbols.

The decision block is used to answer a question. It could involve a quality check, a piece count, or other parameter. The question is generally a yes or no question. The process block is used to perform a process. It could be an arithmetic calculation, an assembly step, or any other logical processing step. The input/output block is used to input information into the system and the output block is used to output system information. The input might be from an operator or from the system. The output might be a printed report or a display on a terminal. Arrows are used to connect the blocks and to show the direction of operation.

Flow diagramming can make an application program development much easier. Remember to break the application into small logical steps. Once the flow diagram is complete, it must be tested. This can be accomplished by applying some sample data and following it through the flow diagram. If the flow diagram works with sample data, the odds are good that your logic is good. Once the flow diagram is complete, it is a rather simple task to code the application into ladder logic.

Remember that a well-planned job is half done.

Flow diagramming helps with proper planning.

Figure 4-28 shows a heat treat application that was flow diagrammed. The sequence of operations is explained below.

> **The operator inputs the number of parts to be hardened.**
>
> **The part is preheated to 1000 degrees Fahrenheit.**
>
> **Next, the part is heated to 1650 degrees for hardening.**
>
> **The part is then quenched in oil to harden it.**
>
> **The hardness is then checked; if it is incorrect, the part is scrapped.**
>
> **Then the piece count is checked to see if enough parts have been hardened; if not, more are hardened.**
>
> **When enough have been made, the process is ended.**

The second technique for developing programs is called pseudocode. Pseudocode is an English representation of the system steps. The use of pseudocode requires the programmer to write logical steps for the system in descriptive statements or blocks. Pseudocode would not look much different than the steps for the flow diagram that was just developed. In fact, they should be the same because the system should operate the same whether flow diagramming is used or pseudocode is used.

The purpose of either method is to help the programmer think logically before the actual programming is begun.

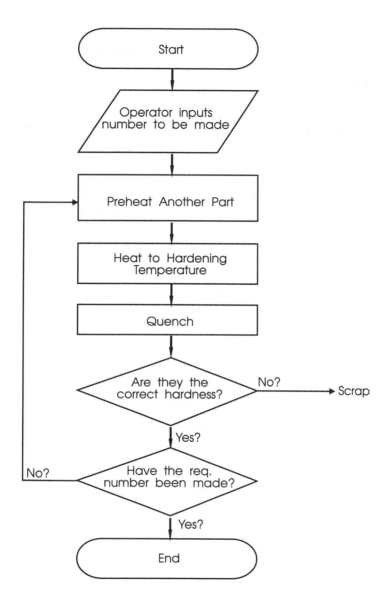

Figure 4-28. This figure shows the flow diagram for a simple hardening process.

Questions

1. What are timers typically used for?
2. Explain the two types of timers and how each might be used.
3. What does the term <u>retentive</u> mean?
4. Draw a typical retentive timer and describe the purpose of the inputs.
5. What are counters typically used for?
6. In what way are counters and timers very similar?
7. Explain the two contacts that are usually required for a counter.
8. What is cascading?
9. You have been asked to program a system that requires that completed parts are counted. The largest counter available in the PLC's instruction set can only count to 999. We must be able to count up to 5000. Draw a ladder diagram that shows the method you would use to complete the task. Hint: use two counters, one as the input to the other. The total count will involve looking at the total of the two counters.
10. Describe at least three reasons for using a flow diagram or pseudocode in program development.
11. Figure P4-11A is a partial drawing of a heat-treat system. You have been asked by your supervisor to troubleshoot the system. The engineer who originally developed the system no longer works for your company. He never fully documented the system. A short description of the system follows. You must study the drawings and data and complete this assignment.

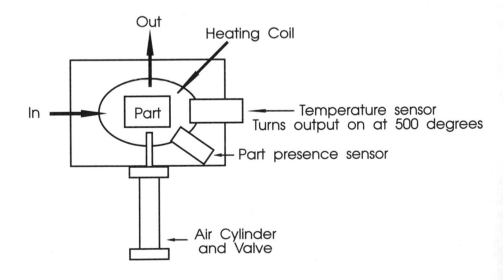

Figure P4-11A

A part enters this portion of the process. The temperature must then be raised from room temperature to 500 degrees Fahrenheit. There is also a part presence sensor. There is also a sensor that turns on when the temperature reaches 500 degrees. The part must then be pushed out of the machine. The cycle should take about 25 seconds. If it takes an excessive amount of time and the temperature has still not been reached, an operator must be informed and must reset the system. Study the system drawing, I/O chart, and ladder diagram, then complete the I/O table and answer the questions.

Complete the I/O chart (Figure P4-11B) by writing short comments that describe the purpose of each input, output, timer, and counter. Make them very clear and descriptive so that the next person to troubleshoot the system would have an easier task. Then refer to Figure P4-11C and answer the following questions.

System I/O	
X3	Part Presence Sensor
X20	Operator Reset
Y5	
Y8	
Timer 1	
Timer 9	
Counter 1	
Counter 2	

Figure P4-11B

Answer the following questions.
a. What is the purpose of counter 1?

b. What is the purpose of input X3?

c. What is input X20 being used for?

d. What is the purpose of counter 2?

e. What is the purpose of contact Y15?

f. Part of the logic is redundant. Identify the part and suggest a change.

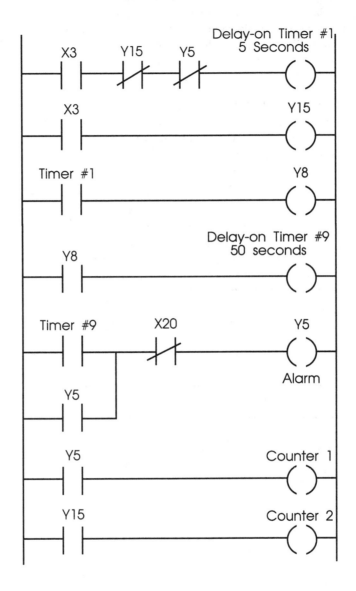

Figure P4-12

12. Examine the ladder diagram in Figure P4-12. Assume that input 00001 is always true. What will this ladder logic do?

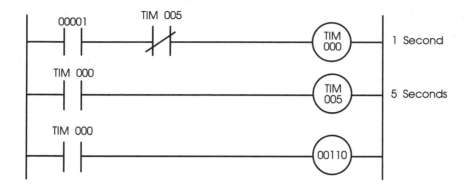

Figure P4-13

13. You have been assigned the task of developing a stoplight application. Your company thinks there is a large market in intelligent street corner control. Your company is going to develop a PLC based system in which lights will adapt their timing to compensate for the traffic volume. Your task is to program a normal stoplight sequence to be used as a comparison to the new system.

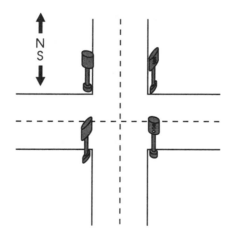

Figure P4-14.

Note in Figure P4-14 that there are really two sets of lights. The north and south lights must react exactly alike. The east-west set must be the complement of the north-south set. Write a program that will keep the green light on for 25 seconds and the yellow for 5 seconds. The red will then be on for 30 seconds. You must also add a counter because the bulbs are replaced at a certain count for preventive maintenance. The counter should count complete cycles. (Hint: To simplify your task, do one small task at a time. Do not try to write the entire application at once. Write ladder logic to get one light working, then the next, then the next, before you even worry about the other stoplight. When you get one set done, the other is a snap. Remember: A well-planned job is half done.)

Develop the flow diagram and then write the ladder diagram. (<u>Hint</u>: It might be helpful to develop one flow diagram for the east-west lights and one for the north-south set.)

Additional Exercises

1. Enter the stoplight program that you developed into TISOFT programming software. The software is available from your instructor. Enter synonyms and comments for all contacts and coils. Print the complete program with synonyms, comments and a cross reference. Review the TISOFT tutorial software if necessary.

Chapter 5

Sensors and Their Wiring

In this chapter we examine types and uses of industrial sensors. Digital and analog sensors are covered, and we also examine the wiring of sensors.

OBJECTIVES

Upon completion of this chapter, the student will be able to:

Describe at least two ways in which sensors can be classified.

Choose an appropriate sensor for a given application.

Describe the typical uses of digital sensors.

Describe the typical uses of analog sensors.

Explain common sensor terminology.

Explain the wiring of load and line-powered sensors.

Explain how field-effect sensors function.

Explain the principle of operation of thermocouples.

Explain such common thermocouple terms as <u>types</u> *and* <u>compensation</u>.

The Need For Sensors

Sensors have become vital in industry. Manufacturers are moving toward integrating pieces of computer-controlled equipment. In the past, operators were the brains of the equipment. The operator was the source of all information about the operation of a process. The operator knew if there were parts available, which parts were ready, if they were good or bad, if the tooling was OK, if the fixture was open or closed, and so on. The operator could sense problems in the operation. He/she could see, hear, feel (vibration, etc.), even smell problems.

Industry is now using computers (in many cases PLCs) to control the motions and sequences of machines. PLCs are much faster and more accurate than an operator at these tasks. PLCs cannot see, hear, feel, smell, or taste processes by themselves. Industrial sensors are used to give the PLC these capabilities.

Simple sensors can be used by the PLC to check if parts are present or absent, to size the parts, even to check if the product is empty or full. The use of sensors to track processes is vital for the success of the manufacturing process and also to assure the safety of the equipment and operator.

Sensors, in fact, perform simple tasks more efficiently and accurately than people do. Sensors are much faster and make far fewer mistakes.

Studies have been done to evaluate how effective human beings are in tedious, repetitive, inspection tasks. One study examined people inspecting table tennis balls. A conveyor line was set up to bring table tennis balls past a person. White balls were considered good, black balls were considered scrap. It was found that people were only about 70 percent effective at finding the defective ping pong balls. Certainly people can find all of the black balls, but they do not perform mundane, tedious, repetitive tasks well. People become bored and make mistakes. Whereas a simple sensor can perform simple tasks almost flawlessly.

Typical Applications

One of the most common uses of a sensor is in a product feeding situation. It may involve parts moving along a conveyor or in some type of parts feeder (See Figure 5-1). The sensor is used to notify the PLC when a part is in position, ready to be used. This is typically called a presence/absence check.

The same sensor can also provide the PLC with additional information. The PLC can take the data from the sensor and use it to count the parts as they are sensed. The PLC can also compare the completed parts and time to compute cycle times.

This one simple sensor allowed the PLC to accomplish three tasks:

> *Are there parts present?*
>
> *How many parts have been used?*
>
> *What is the cycle time for these parts?*

Chapter 5: Sensors and Their Wiring

Simple sensors can be used to decide which product is present. Imagine a manufacturer that produces three different size packages on the same line. Now imagine the product sizes moving along a conveyor at random. When each package arrives at the end of the line, the PLC must know what size product is present.

This can be done very easily with three simple sensors. If only one sensor is on, the small product is present. If two sensors are on, it must be the middle-sized product. If three sensors are on, the product must be the large size. The same information could then be used to track production for all product sizes and cycle times for each size.

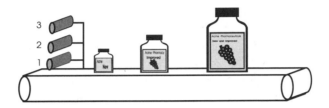

Figure 5-1. Use of sensors to check the size of containers as they move along a conveyor. This information could be used to divert the product to the appropriate processes.

Sensors can even be used to check whether or not containers have been filled. Imagine aspirin bottles moving along a conveyor with the protective foil and the cover on. There are simple sensors that can sense right through the cap and seal and make sure that the bottle was filled. One sensor would be set up to sense when there is a bottle present. This is often called a gate sensor. A gate sensor is used to show when a product is in place. The PLC then knows that a product is present and can perform further checks.

 A second sensor would be set up to sense the aspirin in the bottle. If there is a bottle present, but the sensor does not detect the aspirin inside, the PLC knows that the aspirin bottle was not filled. The PLC can then make sure that no empty bottles leave the plant. The PLC can also track scrap rates, production rates, and cycle times.

Sensors can be used to monitor temperature. Imagine an oven used in a bakery. The sensor can monitor the temperature. The PLC can then control the heater element in the oven to maintain the ideal temperature.

Pressure is vital in many processes. Imagine a plastic injection machine. Heated plastic is forced into a mold under a given pressure. The pressure must be accurately maintained or the parts will be defective. Sensors can be used to monitor these pressures. The PLC can then monitor the sensor and control the pressure.

Flow rates are important in process industries such as papermaking. Sensors can be used to monitor the flow rates of raw material. The PLC can use these data to adjust and control the flow rate of the system. Think about your water supply at home. The water department monitors the flow of water to calculate your bill.

The applications noted above are but a few of the simple uses to which sensors can be put.

The innovative engineer or technician will invent many other uses. You should now be aware that the data from one sensor can be used to provide many different types of information (i.e., presence/absence, part count, cycle time, etc.).

The types of sensors and the complexity of their use in solving application problems grows daily. New sensors are constantly being introduced to solve needs. There are even magazines devoted to the topic of sensors.

Sensor Types

Contact vs. Noncontact

Sensors are classified in a number of ways. One common way is to divide sensors into two categories: contact and noncontact. This is a simple way to identify a sensor. If the device must contact a part to sense the part, it would be called a <u>contact sensor</u>. A simple limit switch on a conveyor is an example. When the part moves the lever on the switch, the switch changes state. The contact of the part and the switch creates a change in state that the PLC can monitor. Noncontact sensors are sensors that can detect the product without touching the product physically.

<u>Noncontact sensors</u> do not operate mechanically (no moving parts). Mechanical devices are much less reliable than electronic devices. This means that noncontact sensors are less likely to fail.

Speed is another consideration. Electronic devices are much faster than mechanical devices. Noncontact devices can perform at very high production rates.

Another advantage of not touching the product is that you do not slow down or interfere with the process.

In the remainder of the chapter we examine noncontact sensors.

Digital vs. Analog

Another way in which sensors can be classified is digital or analog. Digital sensors are the easiest to use. Computers are digital devices. A computer actually works only with ones or zeros (on or off). A digital sensor has two states: on or off. Most applications involve presence/absence and counting. A digital sensor meets this need perfectly and inexpensively.

Analog sensors are more complex but can provide much more information about a process. Analog sensors are often called <u>linear output</u> sensors. Temperatures are analog information.

Think about a sensor used to measure the temperature. The temperature in the Midwest is usually between 0 and 90 degrees. An analog sensor could sense the temperature and send a current to the PLC. The higher the temperature, the higher the output from the sensor. The sensor may, for example, output between 4 and 20 mA, depending on the actual temperature. There are an unlimited number of temperatures (and thus current outputs). Remember that the output from the digital sensor was on or off. The output from the analog sensor can be any value in the range from low to high. Thus the PLC can monitor temperature very accurately and control a process closely. Pressure sensors are also available in analog style. They provide a range of output voltage (or current), depending on the pressure. There are needs for both digital and analog sensors in industrial applications. Digital sensors are more widely used because of their simplicity and ease of use. There are, however, applications that require information that only analog sensors can provide.

Digital Sensors

Digital sensors come in many types and styles. Types of digital sensors are examined next.

Optical Sensors

Optical sensors use light to sense objects. Optical sensors are increasing in use. In the past optical sensors were somewhat unreliable because they used common light and were affected by ambient lighting. This caused intermittent problems and made them somewhat unpopular. The optical sensors of today have solved these problems. Optical sensors today are very reliable because of the way they now operate.

All optical sensors function in approximately the same manner. There is a light source (emitter) and a photodetector to sense the presence or absence of light. Light-emitting diodes (LEDs) are typically used for the light source.

LEDs are chosen because they are small, sturdy, very efficient, and can be turned on and off at extremely high speeds. They operate in a narrow wavelength and are very reliable.

The LEDs in sensors are used in a pulsed mode. The emitter is pulsed (turned off and on repeatedly). The "on time" is extremely small compared to the "off time." LEDs are pulsed for two reasons: so that the sensor is unaffected by ambient light; and to increase the life of the LED. The pulsed light is sensed by the photodetector. The photodetector essentially sorts out all ambient light and looks for the pulsed light. The light sources chosen are typically invisible to the human eye. The wavelengths are chosen so that the sensors are not affected by other lighting in the plant. The use of different wavelengths allows some sensors, called color mark sensors, to differentiate between colors. The pulse method and the wavelength chosen make optical sensors very reliable. All of the optical sensors function in the same basic manner. The differences are in the way in which the light source (emitter) and receiver are packaged.

Light/Dark Sensing
Optical sensors are available in either light or dark sensing. In fact, many sensors can be switched between light and dark modes. Light/dark sensing refers to the normal state of the sensor, whether its output is on or off in its normal state.

Light Sensing
The output is energized (on) when the sensor receives the modulated beam. In other words, the sensor is on when the beam is unobstructed.

Dark Sensing
The output is energized (on) when the sensor does not receive the modulated beam. In other words, the sensor is on when the beam is blocked.

Timing Functions

Timing functions are available on some optical sensors. They are available with on-delay and off-delay. <u>On-delay</u> delays the turning on of the output by a user-selectable amount. <u>Off-delay</u> holds the output on for a user-specified time after the object has moved away from the sensor.

Reflective Sensors

One of the more common types of optical sensors is the reflective or diffuse reflective type. The emitter and receiver are packaged in the same unit. The emitter sends out light which bounces off the product to be sensed. The reflected light returns to the receiver where it is sensed (Figure 5-2). Reflective sensors have less sensing distance (range) than other types of optical sensors because they rely on light reflected off the product.

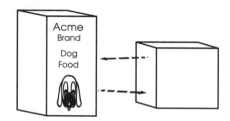

Figure 5-2. Reflective type sensor. The light emitter and receiver are in the same package. When the light from the emitter bounces off an object it is sensed by the receiver and the output of the sensor changes state. The broken-line style of the arrows represents the pulsed mode of lighting, which is used to assure that ambient lighting does not interfere with the application. The sensing distance (range) of this style is limited by how well the object can reflect the light back to the receiver.

Thru-Beam

Another common sensor is the thru-beam (Figure 5-3). In this configuration the emitter and receiver are packaged separately. The emitter sends out light across a space and is sensed by the receiver. If the product passes between the emitter and receiver, it stops the light from hitting the receiver and the sensor knows there is product present.

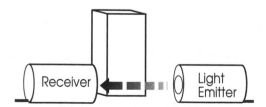

Figure 5-3. Thru-beam sensor. The emitter and receiver are in separate packages. When an object breaks the light path, the output of the receiver portion of the sensor changes state. The broken-line arrow symbolizes the pulsed mode of the light that is used in optical sensors.

Retroreflective

The retroreflective is similar to the reflective sensor (see Figure 5-4). The emitter and receiver are both mounted in the same package. The difference is that the retroreflective sensor bounces the light off a reflector instead of the product. The reflector is similar to the reflectors used on bicycles. Retroreflective sensors have more sensing distance (range) than do reflective (diffuse) sensors but less sensing distance than that of a thru-beam sensor.

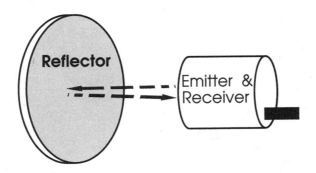

Figure 5-4. Retroreflective sensor. The light emitter and receiver are in the same package. The light bounces off a reflector (similar to the reflector on a bicycle) and is sensed by the receiver. If an object obstructs the beam, the output of the sensor changes state. The excellent reflective characteristics of a reflector give this sensor more sensing range than a typical diffuse style, where the light bounces off the object. The broken-line arrows represent the pulsed method of lighting that is used.

Fiber-Optic Sensors

Fiber-optic sensors are simply mixes of the other types. The actual emitter and receiver are the same. Fiber-optic cables are just attached to both. One cable is attached to the emitter, one to the receiver. These cables are very small and flexible. The light from the emitter passes through the cable and exits from the other end. The light enters the end of the cable attached to the receiver, passes through the cable, and is sensed by the receiver. The cables are available in both thru-beam and reflective configuration. (See Figure 5-5.)

Color Mark Sensors

Color mark sensors are a special type of diffuse reflective optical sensor. This type of sensor can differentiate between colors. They actually do not differentiate between colors. They differentiate between different shades. These are typically used to check labels and for sorting packages by color mark. They are chosen according to the color to be sensed. A sensitivity adjust is provided to make fine adjustments. The background color (behind the object) is an important consideration. Sensor manufacturers have charts available for the proper selection of color mark sensors for various colors.

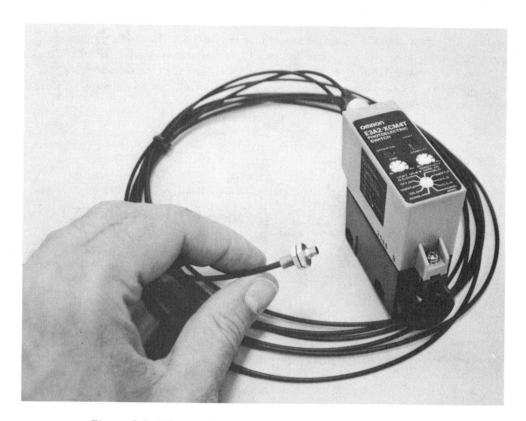

Figure 5-5. Fiber optic sensor. Courtesy of Omron Electronics.

Laser Sensors

Lasers are also used as a light source for optical sensors. These sensors can be used for precision quality inspections. Resolution can be as small as 10 microns. Outputs can be analog or digital. The digital outputs can be used to signal pass/fail, or other indication. The analog output can be used to monitor and record actual measurements.

Ultrasonic Sensors

Ultrasonic sensors use a narrow ultrasonic beam to detect and even measure (see Figure 5-6). The ultrasonic sensor is much like radar. A narrow ultrasonic beam (about 5 mm) is bounced off the object back to the sensor. The sensor is then able to determine the distance to the object. They can be used to detect the size of objects also. Objects as small as 1 mm can be inspected to an accuracy of plus or minus 0.2 mm.

Electronic Field Sensors (Field Sensors)

The two most common types of field sensors function in essentially the same way. They have a field generator and a sensor to sense when the field is interfered with. Imagine the magnetic field from a magnet. The field generator puts out a similar field. The two types of field sensors are inductive and capacitive.

Measuring height of different objects on a conveyor using external gate input to coordinate inspection

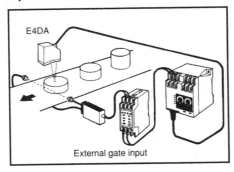

Figure 5-6. Use of an ultrasonic sensor. Note the use of the "gate" sensor to notify the PLC when a part is present. Courtesy of Omron Electronics.

Inductive sensors

The inductive sensor, used to sense metallic (ferrous) objects, is commonly used in the machine tool industry. The inductive sensor works by the principle of electromagnetic induction (see Figure 5-7). Induction sensors function in a manner similar to the primary and secondary windings of a transformer. There is an oscillator and a coil in the sensor. Together they produce a weak magnetic field. As an object enters the sensing field, small eddy currents are induced on the surface of the object. Because of the interference with the magnetic field, energy is drawn away from the oscillator circuit of the sensor. The amplitude of the oscillation decreases, causing a voltage drop. The detector circuit of the sensor senses the voltage drop of the oscillator circuit and responds by changing the state of the sensor.

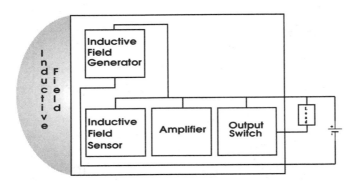

Figure 5-7. Block diagram of an inductive sensor. The inductive field generator creates an inductive field in front of the sensor. This field is monitored by the field sensor. When a ferrous object (an object containing the element iron) enters the field, the field is disrupted. The disruption in the field is sensed by the field sensor and the output of the sensor changes state. The sensing distance of these style sensors is determined by the size of the field. This means that the larger the required sensing range, the larger the diameter of the sensor will be.

Inductive sensors are available in very small sizes. If the area in which the sensor will be mounted is restricted, or if the object to be sensed is small, this style of sensor works very well. (See Figure 5-8.)

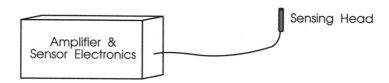

Figure 5-8. Use of a remote inductive sensing head. The electronics are mounted away from the application. The small sensing head is mounted in the application.

Electromagnetic Induction

Refer to Figure 5-9. The output is initially deenergized (off). As the target (object) moves into the field of the sensor, eddy currents are produced along the target surface. As the voltage drops in the oscillator circuit, the detector senses the drop and changes the state of the sensor. The output is then energized.

Sensing Distance

Sensing distance (range) is related to the size of the inductor coil and whether the sensor coil is shielded or nonshielded (see Figure 5-10). This sensor coil is shielded. In this case a copper band is placed around the coil. This prevents the field from extending beyond the sensor diameter. Note that this reduces the sensing distance.

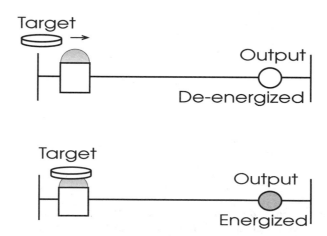

Figure 5-9. How an object (target) is detected as it enters a field.

The shielded sensor has about half the sensing range of an unshielded sensor. Sensing distance is affected by temperature. Sensing distance will generally vary by about 5 percent due to changes in ambient temperature.

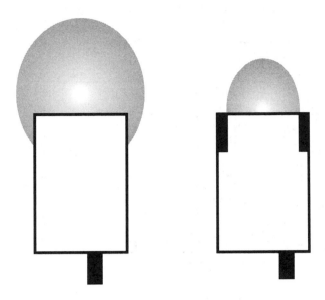

Figure 5-10. This figure shows the use of a copper band in a shielded field sensor. Note that the sensing distance is reduced. It does allow the sensor to be mounted flush, however. If the unshielded sensor were mounted flush, it would detect the object in which it was mounted.

Hysteresis

Hysteresis means that the object must be closer to turn a sensor on than to turn it off (see Figure 5-11). Direction and distance are important. If the object is moving toward the sensor, it will have to move to the closer point. Once the sensor turns on (operation point or on-point), it will remain on until the object moves away to the release point (off-point). This differential gap, or differential travel, is caused by hysteresis. The principle is used to eliminate the possibility of "teasing" the sensor. The sensor is either on or off. (See Figure 5-10.)

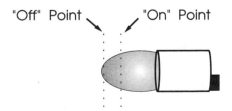

"Off" Point   "On" Point

Figure 5-11. Example of hysteresis. Note that the on-point and off-point are different.

Capacitive Sensors

Capacitive sensors (Figures 5-12 and 5-13) can be used to sense both metallic and nonmetallic objects. They are commonly used in the food industry. Capacitive sensors can also be used to sense for product inside nonmetallic containers (see Figure 5-14).

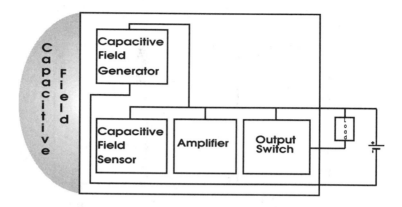

Figure 5-12. Block diagram of a capacitive sensor.

Capacitive sensors operate on the principle of electrostatic capacitance. They function in a manner similar to the plates of a capacitor. The oscillator and electrode produce an electrostatic field. (Remember that the inductive sensor produced an electromagnetic field. See Figure 5-12.)

The target (object to be sensed) acts as the second plate of a capacitor. An electric field is produced between the target and the sensor. As the amplitude of the oscillation increases, the oscillator circuit voltage increases and the detector circuit responds by changing the state of the sensor.

Almost any object can be sensed by a capacitive sensor. The object acts like a capacitor to ground. When the target (object) enters the electrostatic field, the dc balance of the sensor circuit is disturbed. This starts the electrode circuit oscillation and maintains the oscillation as long as the target is within the field.

Sensing Distance

Capacitive sensors are unshielded, nonembeddable devices. This means that they cannot be installed flush in a mount because they would then sense the mount. Conducting materials can be sensed farther away than nonconductors because the electrons in conductors are freer to move. The target mass affects the sensing distance: the larger the mass, the larger the sensing distance.

Capacitive sensors are more sensitive than inductive sensors to temperature and humidity fluctuation. Sensing distance can fluctuate as much as plus or minus 15 to 20 percent. Capacitive sensors are not as accurate as inductive sensors. Repeat accuracy can vary by 10 to 15 percent in capacitive sensors.

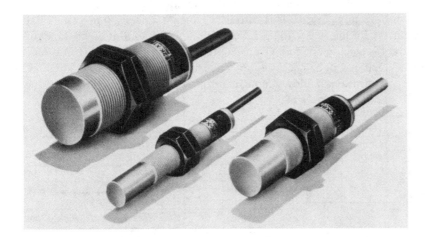

Figure 5-13. Capacitive sensors. Courtesy of Omron Electronics.

Some capacitive sensors are available with a sensitivity adjustment. This can be used to sense product inside a container. The sensitivity can be reduced so that the container is not sensed but the product inside is.

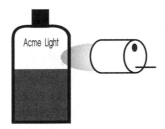

Figure 5-14. Use of a capacitive sensor to check inside a container. They are also used to check fluid and solid levels inside tanks. There is an adjustment screw on some capacitive sensors. The sensor would be adjusted so that the container is not sensed but the material inside is.

Sensor Wiring

There are basically two wiring schemes for sensors: load-powered and line-powered. This applies to ac and dc-powered sensors.

Load-Powered Sensors

The load-powered type is a two-wire sensor (see Figure 5-15). There are only two wires to connect the sensor. The current required for the sensor to operate must pass through the load. A load is anything that will limit the current of the sensor output. Think of the load as being an input to a PLC. The small current flow that must flow to allow the sensor to operate is called leakage current, or operating current. This current is typically under 2.0 mA (see Figure 5-16). This is enough current for the sensor to operate, but not enough to turn on the input of the PLC. (The leakage current is usually not enough current to activate a PLC input. If it is enough current to turn on the PLC input it will be necessary to connect a bleeder resistor as shown in Figure 5-17.) When the sensor turns on, it allows enough current to flow to turn on the PLC input.

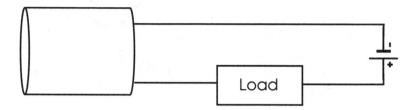

Figure 5-15. Two-wire sensor (load-powered). The load represents whatever the technician will be monitoring the sensor output with. It is normally a PLC input. The load must limit the current to an acceptable level for the sensor, or the sensor output will be blown.

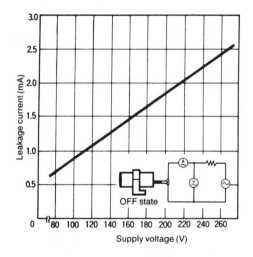

Figure 5-16. Leakage current vs. supply voltage.

Response time is the lapsed time between the target being sensed and the output changing state. Response time can be crucial in high production applications. Sensor specification sheets will give response times.

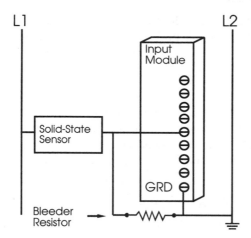

Figure 5-17. Use of a two-wire sensor (load-powered). In this case the leakage current was enough to cause the input module to sense an input when there was none. A resistor was added to bleed the leakage current to ground so that the input could not sense it.

Line-Powered Sensors

Line-powered sensors are usually the three-wire type (see Figure 5-18), but there can be either three or fours. There are two power leads and one output lead in the three-wire variety.

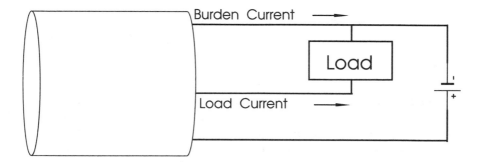

Figure 5-18. Three wire (line-powered) sensor. Note the load. The load must limit the current to an acceptable level.

The sensor needs a small current, called burden current or operating current, to operate. This current flows whether or not the sensor output is on or off. The load current is the output from the sensor. If the sensor is on, there is load current. This load current turns the load (PLC input) on. The maximum load current is typically between 50 to 200 mA for most sensors. Make sure that you limit the load (output) current or the sensor will be ruined.

PNP Sensor (Sourcing Type)

Conventional current flow goes from plus to minus. When the sensor is off, the current does not flow through the load. In the case of a PNP, when there is an output current from the sensor, the sensor sources current to the load (sourcing type).

NPN (Sinking Type)

When the sensor is off (nonconducting), there is no current flow through the load. When the sensor is conducting, there is a load current flowing from the load to the sensor. The choice of whether to use a NPN or a PNP sensor is dependent on the type of load. In other words, choose a sensor that matches the PLC input module requirements for sinking or sourcing.

Other Transducers

Thermocouples

The thermocouple is one of the most common devices for temperature measurement in industrial applications. A thermocouple is a very simple device: two pieces of dissimilar metal wire joined at one or both ends (See Figure 5-19). The typical industrial thermocouple is joined at one end. The other ends of the wire are connected via compensating wire to the analog inputs of a control device such as a PLC (see Figure 5-20). The principle of operation is that when dissimilar metals are joined, a small voltage is produced. The voltage output is proportional to the difference in temperature between the cold and hot junctions (see Figures 5-21 and 5-22).

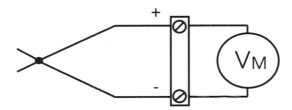

Figure 5-19. Common thermocouple connection. Note that the actual thermocouple wires are only connected to each other on one end. The other end of each is connected to copper wire, which in this case is connected to a meter. They are normally connected at a terminal strip. The terminal strip assures that both ends remain at the same temperature (ambient temperature). The net loop voltage remains the same as the double-ended loop.

The cold junction is assumed to be at ambient temperature (room temperature). Industrial thermocouple tables use 75 degrees Fahrenheit for the reference temperature (See Figure 5-23).

In reality, temperatures vary considerably in an industrial environment. If the cold junction varies with the ambient temperature, the readings will be inaccurate. This would be unacceptable in most industrial applications. It is too complicated to try and maintain the cold junction at

75 degrees. Industrial thermocouples must therefore be compensated. This is normally accomplished with the use of resistor networks that are temperature sensitive. The resistors that are used have a negative coefficient of resistance. Resistance decreases as the temperature increases. This adjusts the voltage automatically so that readings remain accurate. PLC thermocouple modules automatically compensate for temperature variation.

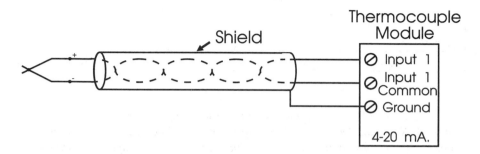

Figure 5-20. Thermocouple. Note that the wire connecting the thermocouple to the PLC module is twisted pair (two wires twisted around each other) shielded cable. The shield around the twisted-pair is to help eliminate electrical noise as a problem. Twisted-pair wiring also helps reduce the effects of electrical noise. Note also that the shielding is grounded only at the control device. Typical PLC modules would allow four or more analog inputs.

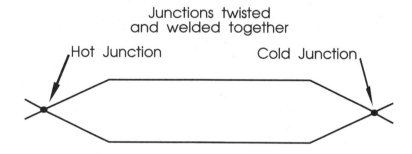

Figure 5-21. Simple thermocouple. A thermocouple can be made by twisting the desired type of wire together and silver soldering the end. To measure the temperature change the wire would be cut in the middle and a meter would be inserted. The voltage would be proportional to the difference in temperature between the hot and cold junctions.

The thermocouple is a very accurate device. The resolution is determined by the device that takes the output from the thermocouple. The device is normally a PLC analog module. The typical resolution of an industrial analog module is 12 bits. 2 to the twelfth power is 4096. This means that if the range of temperature to be measured was 1200 degrees the resolution would be 0.29296875 degrees/bit (1200/4096 = 0.29296875). Which would mean that our PLC could tell the temperature to about one-fourth of one degree. This is reasonably good resolution, close enough for the vast majority of applications. Higher-resolution analog modules are available.

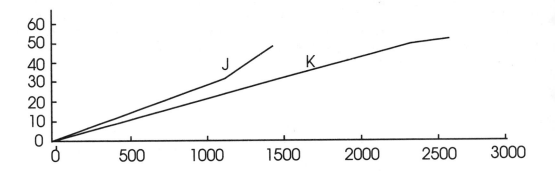

Figure 5-22. Voltage output vs. temperature for J- and K-type thermo-couples. Note that the relationship is approximately linear. For example, if the output voltage was 20 mV with a J-type thermocouple the temperature would be approximately 525 degrees (600-75). Note that the chart shows the voltage produced by the difference in temperature between the cold and hot junctions. The chart assumes a 75 degree ambient temperature. The temperatures shown are Fahrenheit.

Thermocouple Temperature Ranges

Type	Temperature Fahrenheit	Temperature Celsius
J	-50 to +1400	-40 to +760
K	-100 to +2250	-80 to +1240
T	-200 to +660	-130 to +350
E	-200 to +1250	-130 to +680
R	0 to +3200	-20 to +1760
S	0 to +3200	-20 to +1760
B	+500 to +3200	+260 to +1760
C	0 to +4200	-20 to +2320

Figure 5-23. Temperature ranges for various types of thermocouples. Note the output for the difference in temperature between the hot and cold junctions. The table is based on a cold junction temperature of 75 degrees Fahrenheit.

Thermocouples are the most widely used temperature sensors. They are a wide variety of thermocouples available (see Figure 5-24). They are not generally as accurate as thermistors or RTDs.

Thermocouple Types	
E	Nickel-Chromium vs. Copper-Nickel (Chromel-Constantan)
J	Iron vs. Copper-Nickel (Iron-Constantan)
K	Nickel-Chromium vs. Nickel-Aluminum (Chromel-Alumel)
T	Copper vs. Copper - Nickel (Copper-Constantan)
C	Tunsten - 5% Rhenium vs. Tungsten - 26% Rhenium
D	Tungsten vs. Tungsten - 26% Rhenium
G	Tungsten - 3% Rhenium vs. Tungsten - 25% Rhenium
R	Platinum vs. Platinum - 13% Rhodium
S	Platinum vs. Platinum - 10% Rhodium
B	Platinum - 6% Rhodium vs. Platinum - 30% Rhodium

Figure 5-24. Composition for various thermocouple types.

RTDs

An RTD (resistance temperature detector) is a device that changes resistance with temperature. Most RTDs are made with platinum, a very stable element. The change in resistance vs. temperature change is very linear. This makes the RTD a very accurate device. RTDs are connected like resistors. The most common resistance for an RTD is 100 ohms at zero degrees C. Other RTDs are available in the 50 to 1000 ohm range.

Thermistors

Thermistors offer a large change in resistance for a given change in temperature. They can be very precise and stable. The major problem is that thermistors are very nonlinear. If the range of temperature to be measured is relatively small, the thermistor is a good device. Thermistor networks are available that have very linear voltage change with temperature change.

Strain Gages

Strain gages are used to measure force. They are based on the principle that the thinner the wire, the higher the resistance. Strain gages are made of wire that is somewhat elastic. As it is stretched it becomes thinner, thus increasing the wire's resistance. This change in resistance can be measured and converted to pressure or force. Strain gages are normally bonded to a membrane. The pressure or force distorts the membrane and stretches the strain gage. Strain gages are normally used in bridge configurations.

Installation Considerations

Electrical

The main consideration for sensors is to limit the load current. The output (load) current must be limited on most sensors to a very small output current. The output limit is typically between 50 and 200 mA. If the load draws more than the sensor current limit, the sensor fails and you buy a new one. More sensors probably fail because of improper wiring than actually fail from use. It is crucial that you carefully limit current to a level that the sensor can handle (see Figure 5-25). PLC input modules limit the current to acceptable levels. Some sensors are available with relay outputs. These can handle higher load currents (typically 3 A).

Figure 5-25. How <u>not</u> to hook up a two-wire sensor. Remember, do not connect any sensor without a load that will limit the current to an acceptable level, or you will ruin the sensor.

If high-voltage wiring is run in close proximity to sensor cabling, the cabling should be run in metal conduit to prevent the sensor from false sensing, malfunction, or damage.

The other main consideration is to buy the proper polarity sensors. If the PLC module requires sinking devices, make sure that sinking devices are purchased.

Mechanical

Sensors should be mounted horizontally whenever possible (see Figure 5-26). This prevents the buildup of chips and debris on the sensor that could cause misreads.

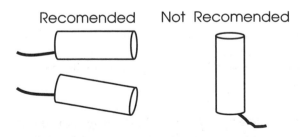

Recomended   Not Recomended

Figure 5-26. Sensor installation. A horizontal mount helps minimize chips and debris from collecting on the sensor and possibly causing misreads. Courtesy of Omron Electronics.

In a vertical position, chips, dirt, oil, and so on, can gather on the sensing surface and cause the sensor to malfunction. In a horizontal position, the chips fall away. If the sensor must be mounted vertically, provision must be made to remove chips and dirt periodically. Air blasts or oil baths can be used.

Care should also be exercised that the sensor does not detect its own mount. For example, an inductive sensor mounted improperly in a steel fixture might sense the fixture. You must also be sure that two sensors do not mutually interfere. If two sensors are mounted too close together, they can interfere with each other and may cause erratic sensing.

You must also be careful not to use too much force when installing a sensor. Many sensor cases are plastic and can be damaged if deformed or impacted by a mounting or holding screw (see Figure 5-27). This deformation can easily ruin a sensor.

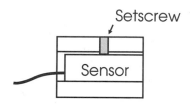

Figure 5-27. How a setscrew is used to hold a sensor in a mount. Care must be exercised that the setscrew is not overtightened. This could easily damage the sensor.

Typical Sensor Applications

When choosing the sensor to use for a particular application there are several important considerations. The characteristics of the object to be sensed are crucial. Is the material plastic? Is it metallic? Is it a ferrous metal? Is the object clear, transparent, or reflective? Is it large or very small?

The specifics of the physical application are very important. Is there a large area available to mount the sensor. Are there problems with contaminants? What speed of response is required? What sensing distance is required? Is excessive electrical noise present? What accuracy is required?

Answering these questions will help narrow down the available choices. A sensor must then be chosen based on such criteria as the cost of the sensor, the cost of failure, and reliability.

Physical Arrangement

The physical arrangement of sensors can be used to differentiate between products (see Figure 5-28). Normally, one sensor is used as a "gate" sensor. A gate sensor is used as a trigger to the PLC. It is used to tell the PLC that a product is in position and further actions should take place. In the case of Figure 5-28, when the gate sensor triggers the PLC, the PLC looks at the state of the two other sensors. If both top sensors are on, it means that the product must be a square. If only the right sensor is on, it means that the product must be a circle. If the gate sensor is on and neither top sensor is on, the product must be a triangle.

There are two important concepts here. The first is that the physical characteristics of an object can be sensed by the innovative use of sensors. Shape can be checked. Sensors can check for holes, protruding surfaces, and to see if parts have been attached to the product.

The second important concept is the use of a sensor as a gate. I have seen a manufacturer scrap a very sophisticated PLC control system, complete with radio-frequency scanning, because it exhibited erratic performance. The erratic performance was due to the fact that the product was being scanned for characteristics without a gate sensor to choose the correct time to scan. The lack of a gate sensor meant that the PLC sometimes scanned the product ahead or behind the desired product. This company scrapped a multithousand-dollar system when a 30 dollar sensor could have solved the problem. The gate sensor could have assured that the proper product was scanned.

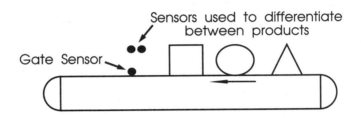

Figure 5-28. How the position of sensors can be used to determine which product is present.

The physical characteristics of the sensor are also important. Special shapes and configurations are available. Horseshoe-shaped optical sensors are available to sense objects passing through the horseshoe. Ring sensors are available to sense objects passing through the ring. Miniature sensing heads are available when space is at a premium or when the object to be sensed is small. Flat, very thin sensor packages are available when space is at a minimum. The point is that there are sensors available to meet any need. A technician should take time to glance through sensor catalogs to see what is available.

As systems become more automated they reduce the opportunity for human observation and intervention. When an operator drills holes in an engine block he/she knows immediately if the drill breaks. The operator then stops the process and replaces the drill. When the process is automated, there is no operator to observe problems. The PLC could care less if the drill breaks. Sensors must be used to sense problems in an automated system. The cost of failure is usually the guide to how much sensing must be done. If the cost is high, sensors should be used to notify the PLC.

Sensors can be used to check whether product has been correctly assembled. Figure 5-30 shows a small inductive sensor being used to check for correct assembly. Sensors can be used to check whether or not machining operations have been performed. Sensors can also be used to check for the presence/absence of electrical components. In Figure 5-31 sensors are being used to sort packages.

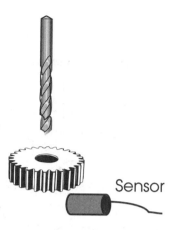

Figure 5-29. Use of a sensor to make sure that the drill actually drills through the gear. Note that the sensor is mounted sideways to allow the chips to fall through.

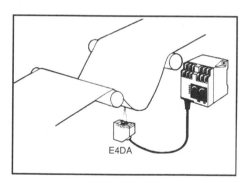
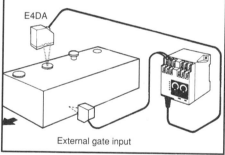

Figure 5-30. Two typical applications, in which ultrasonic sensors are used to measure distances. The application on the left is using the analog output of the sensor for precise control of the web. The application on the right is measuring the height of objects. The fiber optic is used as a gate to indicate part presence. The ultrasonic sensor then takes a reading and the height of the part is known. Courtesy of Omron Electronics.

There are almost as many sensor types available as there are applications. The innovative use of sensors can help increase the safety, reliability, productivity, and quality of processes. Sensors are crucial to the future of American manufacturing. The technician must be able to choose, install, and troubleshoot sensors properly.

Sorting packages by color mark and size

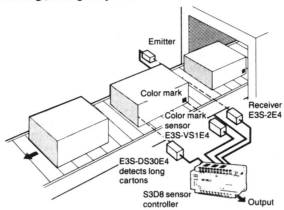

Figure 5-31. Sorting packages with color mark sensors. Courtesy of Omron Electronics.

Questions

1. Describe at least four uses of digital sensors.

2. Describe at least three analog sensors and their use.

3. List and explain at least four types of optical sensors.

4. Explain how capacitive sensors work.

5. Explain how inductive sensors work.

6. Explain the term <u>hysteresis</u>.

7. Draw and explain the wiring of a load-powered sensor.

8. Draw and explain the wiring of a line-powered sensor.

9. What is burden current? Load current?

10. Why must load current be limited?

11. Explain the basic principle on which a thermocouple is based.

12. How are changes in ambient temperature compensated for?

13. What is the temperature range for a type J thermocouple?

14. Explain at least three electrical precautions as they relate to sensor installation.

15. Explain at least three mechanical precautions as they relate to sensor installation.

Chapter 6

Input/Output Modules and Wiring

Originally, PLCs were designed strictly for simple digital (on/off) control. Over the years, PLC manufacturers have added to the capabilities of the PLC. Today, there are I/O cards available for almost any application imaginable.

OBJECTIVES

Upon completion of this chapter, the student will be able to:

List at least two I/O cards that can be used for communication.

Describe at least five special-purpose I/O cards.

Define such terms as <u>resolution</u>, <u>high density</u>, <u>discrete</u>, <u>TTL</u>, and <u>RF</u>.

Choose an appropriate I/O module for a given application.

I/O Modules

Originally, PLCs were used to control one simple machine or process. The changes in American manufacturing have required much more capability. The increasing speed of production and the demand for higher quality require closer control of industrial processes. PLC manufacturers have added modules to meet these new requirements.

Special modules have been developed to meet almost any imaginable need. Modules have been developed to control processes closely. Temperature control is one example.

Industry is beginning to integrate its equipment so that data can be shared. Modules have been developed to allow the PLC to communicate with other devices, such as computers, robots, and machines.

Velocity and position control modules have been developed to meet the needs of accurate high-speed machining. These modules also make it possible for entrepreneurs to start new businesses that design and produce special-purpose manufacturing devices, such as packaging equipment, palletizing equipment, and various other production machinery.

These modules are also designed to be easy to use. They are intended to make it easier for the engineer to build an application. In the balance of this chapter we examine many of the modules that are available.

Digital (Discrete) Modules

Digital modules are also called discrete modules because they are either on or off. A large percentage of manufacturing control can be accomplished through on/off control. Discrete control is easy and inexpensive to implement.

Digital Input Modules

These modules accept an "on" or "off" state from the real-world. The module inputs are attached to devices such as switches or digital sensors. The modules must be able to buffer the CPU from the real-world. Assume that the input is 250 V ac. The input module must change the 250 V ac level to a low-level dc logic level for the CPU. The modules must also optically isolate the real-world from the CPU. Input modules usually have fuses for module protection.

Input modules typically have light emitting diodes (LEDs) for monitoring the inputs. There is one LED for every input. If the input is on, the LED is on. Some modules also have fault indicators. The fault LED turns on if there is a problem with the module. The LEDs on the modules are very useful for troubleshooting.

Most modules also have plug-on wiring terminal strips. All wiring is connected to the terminal strip. The terminal strip is plugged onto the actual module. If there is a problem with a module the entire strip is removed, a new module inserted, and the terminal strip plugged into the new module. Note that there is no rewiring. A module can be changed in minutes or less. This is vital considering the huge cost of a system being down. (The term down means "unable to produce product.")

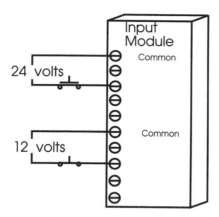

Figure 6-1. Input module with dual commons. This allows the user to mix input voltages.

Input modules usually need to be supplied with power. (Some of the small PLCs are the exception.) The power must be supplied to a common terminal on the module, through an input device, and back to a specific input on the module. Some modules provide multiple commons. This allows the user to mix voltages on the same module (see Figure 6-1). These commons can be jumpered together if desired (see Figure 6-2).

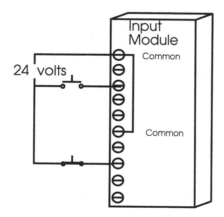

Figure 6-2. How dual commons can be wired together. All inputs would use the same voltage.

When load-powered sensors are used there is always a small leakage current that is necessary for the operation of the sensor. This is not normally a problem. In some cases, however, this leakage current is enough to trigger the input of the PLC module. In this case a resistor can be added that will "bleed" the leakage current to ground (see Figure 6-3). When a bleeder resistor is added, most of the current goes through it to common. This assures that the PLC input turns on only when the sensor is really on.

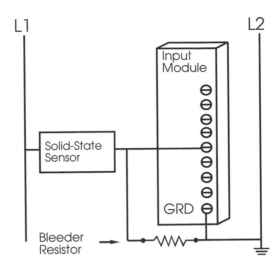

Figure 6-3. Use of a "bleeder" resistor.

High-Speed Counter Modules

High speed counter modules are used to count pulses from sensors, encoders, switches, and so on, at very high speeds. Modules to count pulses up to 75 kHz are common. These units are designed to make the engineer's job simpler. The counter module counts and accumulates the high-speed pulses. The PLC can do its other tasks and check on the count as it needs to. These are really high-speed special-purpose, digital input modules.

Digital Output Modules

Discrete output modules are used to turn real-world output devices either on or off. Discrete output modules can be used to control any two-state device. Output modules are available in ac and dc versions. They are available in various voltage ranges and current capabilities.

The current specifications for a module are normally given as an overall module current and as individual output current. The specification may rate each output at 1 A. If it is an eight-output module, one might assume that the overall current limit would be 8 A (8X1). This is normally not the case. The overall current limit will probably be less than the sum of the individuals. For example, the overall current limit for the module might be 5 A. The user must be careful that the total current to be demanded does not exceed the total that the module can handle. Normally the user's outputs will not each draw the maximum current, nor will they normally all be on at the same time. The user should consider the worst case when choosing an appropriate output module.

Output modules are normally fused to protect each output. There will be a fuse for each output. Many modules also have lights to indicate when fuses are burned out. The overall module is normally fused to prevent damage in case the overall current limit is exceeded. The fuses are normally very easily replaced. The wiring harness is removed, the module is removed from the rack, and the fuse is replaced. Check the technical manual for the PLC to find the exact procedure.

Some output modules provide more than one common terminal. This allows the user to use different voltage ranges on the same card (see Figure 6-4). These multiple commons can be tied together if the user desires. All outputs would be required to use the same voltage, however (see Figure 6-5).

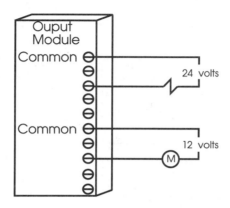

Figure 6-4. Use of a dual-common output module.

Output modules can be purchased with transistor output, triac output, or relay output. The transistor output would be used for dc outputs. There are various voltage ranges and current ranges available, as well as TTL (transistor-transistor logic) output. Triac outputs are used for ac devices. They are also available in various voltage ranges and current ranges. The relay outputs are found quite often on small PLCs. Relay outputs can be used with ac or dc voltages. The voltages can even be mixed (see Figure 6-6). Relay output modules are also available for larger PLCs.

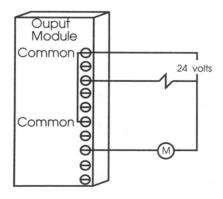

Figure 6-5. Use of a dual-common output module with the commons tied together. Note that the voltages must be the same.

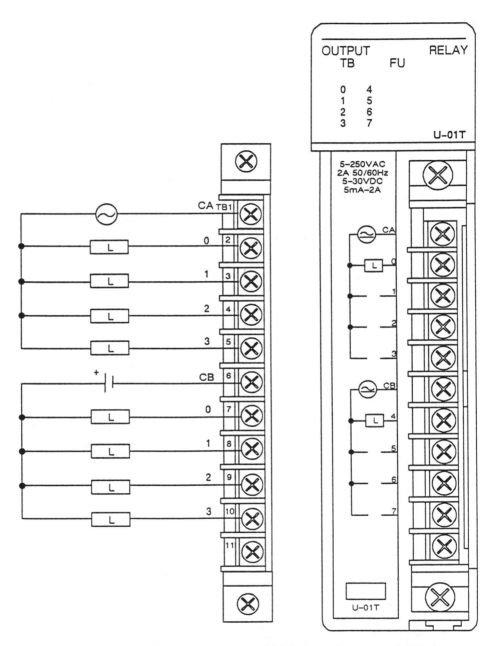

Figure 6-6. Wiring of a relay output module for a Siemens Industrial Automation, Inc. module. Note that ac and dc have been mixed on the same output module. Each output device has been labeled with an "L." The "L" stands for load. Courtesy of Siemens Industrial Automation, Inc..

Chapter 6: Input/Output Modules and Wiring

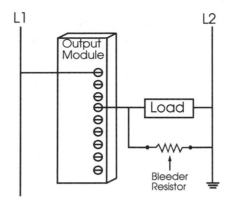

Figure 6-7. Use of a bleeder resistor to "bleed off" unwanted leakage through an output module. This leakage can occur with the solid-state devices that are used in output modules. This is especially true when the load is a high-impedance device.

High-Density I/O Modules

High-density modules are digital I/O modules. A normal I/O module has eight inputs or outputs. A high-density module may have up to 32 inputs or outputs. The advantage is that there are a limited number of slots in a PLC rack. Each module uses a slot. With the high-density module, it is possible to install 32 inputs or outputs in one slot. The only disadvantage is that the high-density output modules cannot handle as much current per output.

Analog Modules

Computers (PLCs) are digital devices. They do not work with analog information. Analog data such as temperature must be converted to digital information before the computer can work with it.

Analog Input Modules

Cards have been developed to take analog information and convert it to digital information. These are called analog-to-digital (A/D) input cards. There are two basic types available: current sensing and voltage sensing. These cards will take the output from analog sensors (such as thermocouples) and change it to digital data for the PLC.

Voltage input modules are available in two types: unipolar and bipolar. Unipolar modules can take only one polarity for input. For example, if the application requires the card to measure only 0 to +10 V, a unidirectional card will work. The bipolar card will take input of positive and negative polarity. For example, if the application produces a voltage between -10 V and +10 V, a bidirectional input card is required, that is because the measured voltage could be negative or positive. Analog input modules are commonly available in 0 to 10 volt models for the unipolar and -10 to +10 V for the bipolar model.

Analog models are also available to measure current. These typically measure 4 to 20 mA.

Many analog modules can be configured by the user. Dip switches or jumpers are used to configure the module to accommodate different voltages or current. Some manufacturers make modules that will accept voltage or current for input. The user simply wires to either the voltage or the current terminals, depending on the application.

Resolution in Analog Modules

Resolution can be thought of as how closely a quantity can be measured. Imagine a 1-ft. ruler. If the only graduations on the ruler were inches, the resolution would be 1 in. If the graduations were every 1/4 in., the resolution would be a 1/4 in. The closest we would be able to measure any object would be l/4 inch.

That is the basis for the measure of an analog signal. The computer can only work with digital information. The analog-to-digital (A/D) card changes the analog source into discrete steps.

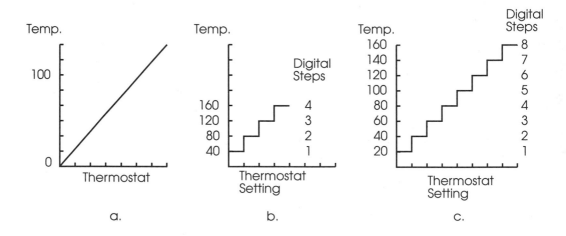

Figure 6-8. Graphs of temperature vs. thermostat setting. Note that graph (a) represents a linear relationship of temperature vs. setting. In reality, when analog control is used, the analog is really a series of steps (resolution). Graph (b) shows what a four-step system would look like. The resolution would be 40 degrees per step. Graph (c) shows an eight-step system. The resolution of graph (c) is 20 degrees.

Examine Figure 6-8a. Ideally, the PLC would be able to read an exact temperature for every setting of the thermostat. Unfortunately, PLCs work only digitally. Consider Figure 6-8b. The analog input card changes the analog voltage (temperature) into digital steps. In this example, the analog card changed the temperature from 40 degrees to 160 degrees in four steps. The PLC would read a number between one and four from the A/D card. A simple math statement in the ladder could change the number into a temperature. For example, assume that the temperature was 120 degrees. The A/D card would output the number 3. The math statement in the PLC would take the number and multiply by 40 to get the temperature. In this case, it would be 3X40, or 120 degrees. If the PLC read 4 from the A/D card, the temperature would be 4X40 or 160. Assume now that the temperature is 97. The A/D card would output the number 2. The PLC would read 2 and multiply by 40. The PLC would believe the temperature to be 80 degrees. The closest the PLC can read the temperature is about 20 degrees if four steps are used. (The temperature that the PLC calculates will always be in a range from 20 degrees below the actual temperature to 20 degrees above the measured temperature.) Each step is 40 degrees. This is called <u>resolution</u>. The smallest temperature increment is 40 degrees. The resolution would be 40.

Consider Figure 6-8c. There are eight steps on this A/D card. The resolution would be twice as fine or 20 degrees. For a temperature of 67 degrees, the A/D would output 3. The PLC would multiply 3X20 and assume the temperature to be 60. The largest possible error would be approximately 10 degrees. (The PLC calculated temperature would be within 10 degrees below the actual temperature to 10 degrees above the actual temperature.)

Industry requires much finer resolution. Typically, an industrial A/D card for a PLC would have 12-bit binary resolution, which means that there would be 4,096 steps. In other words, the analog quantity to be measured would be broken into 4096 steps. Very fine resolution! There are cards available with even finer resolution than this. The typical A/D card is 12 bit binary (4096 steps).

Analog modules are available that can take between one and eight individual analog inputs. Special-purpose A/D modules are also available. One example would be thermocouple modules. These are just A/D modules that have been adapted to meet the needs of thermocouple input. Thermocouples output very small voltages. To be able to use the entire range of the module resolution a thermocouple module amplifies the small output from the thermocouple so that the entire 12-bit resolution is used. The modules also provide cold junction compensation.

These modules are available to make it easy to accept input from thermocouples. The modules are available for various types of thermocouples.

Analog Output Modules

Analog output modules are also available. The PLC works in digital, so the PLC outputs a digital number (STEP) to the digital-to-analog (D/A) converter module. The D/A converts the digital number from the PLC to an analog output. The analog output is typically a voltage between -10 to +10 V dc.

Imagine a bakery. A temperature sensor (analog) in the oven could be connected to an A/D input module in the PLC. The PLC could read the voltage (steps) from the A/D card. The PLC would then know the temperature. The PLC can then send digital data to the D/A output module, which would control the heating element in the oven. This would create a fully integrated, closed-loop system to control the temperature in the oven.

Remote I/O Modules

Special modules are available for some PLCs that allow the I/O module to be positioned separately from the PLC. In some processes it is desirable (or necessary) to position the I/O at a different location. In some cases the machine or application is spread over a wide physical area. In these cases it may be desirable to position the I/O modules away from the PLC. There are many ways to link the I/O module to the PLC. Two common methods are twisted pair (wire) and fiber optics. Twisted pair is the cheaper method. Two wires are twisted around each other and connected between the PLC and the remote I/O. Twisting reduces the possibility of electrical interference (noise). Twisted-pair connections can transmit data thousands of feet.

Fiber-optic links are also available. Fiber-optic cables are noise immune because the data is transmitted as light. Much higher speeds are possible with fiber-optics. The cost of fiber-optics is rapidly decreasing. Fiber-optics can be used to transmit over distances of miles. Remote I/O modules may be purchased for either type of transmission medium.

Communication Modules

Communications are becoming a much larger part of the PLC's task. As systems are integrated, data must be shared throughout the system. PLCs must be able to communicate with computers, CNC machines, robots, and even other PLCs. Much of this communication involves file or data transfer. For example, in a flexible system, a PLC might pass a program to a CNC machine. The CNC machine receives the program and then runs the program. The PLC was not originally designed to perform these duties. Various communication modules are now available to meet most of these needs.

The two main types are host-link and peer-to-peer modules.

Host-Link Modules

One type of communication module is called a host-link module. The host-link module is used to communicate with a host. The host could be a computer or another PLC (see Figure 6-9). Most PLCs have computer software available so that a computer can be used to program the PLC.

A typical use of a host-link module might be an integrated cell. Imagine a variable in the PLC ladder. The variable might contain the number of pieces to be produced in the cell. Now imagine a central computer (host) that would write a number (the number of pieces to be produced) into that variable in the PLC (download). The host-link module makes this possible. RS-232 (serial) communication devices are normally used.

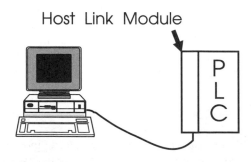

Figure 6-9. Use of a host-link module.

The host link can also be used to send data from the PLC to the computer (upload). The computer can then track process times, completed parts, and so on.

Peer-to-Peer Modules

Peer-to-peer modules are used by PLCs to communicate with other PLCs of the same brand (see Figure 6-10). This is usually done by assigning each PLC a different number. For example, some manufacturers allow 256 PLCs to be on the same network. Each PLC has a unique number (name) between 0 and 255. They communicate over a proprietary network.

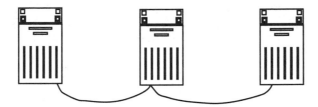

Figure 6-10. Peer-to-peer communications.

ASCII Modules

There are modules available that will transmit and receive ASCII files. These ASCII files are usually programs or manufacturing data. These modules are normally programmed with basic commands. The user writes a program in the basic language. The basic program works with the ladder logic as the program runs.

The modules are designed to make it easier for the PLC user to communicate with various devices. These modules typically have 2 to 24 kilobytes of memory. These modules can also be used to create an operator interface. Basic modules can be used to output text to a printer or terminal in order to update an operator. Communication can be done in several ways.

Position Control Modules

There are modules available for open- and closed-loop position control. Closed-loop systems have feedback devices to ensure that the command is completed.

Open-Loop Position Control

These are modules available to drive stepper motors (see Figure 6-11). Stepper motors can be used to control position in low-power, low-speed applications. They can be used to control positioning of axes very precisely. These modules are designed to make stepper motor integration easy. There are typically acceleration and deceleration functions, as well as teach functions, available. There are also inputs on the module that can be used to home the motor and for overtravel protection.

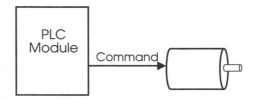

Figure 6-11. Block diagram of an open-loop servo. The PLC module issues a command to the motor and assumes it follows the command. There is no feedback. This is a very common configuration with stepper motors.

Closed-Loop Position Control

There are many complex control applications. Robots and CNC machines are typical examples of closed-loop position control applications. The applications typically involve ac or dc motors to drive tables or axis of motion. The position loop is typically closed (monitored) by encoders. The velocity loop is normally closed by a tachometer. The modules are available to monitor and control velocity and position.

Study Figure 6-12. In this case there are two closed loops. There is a velocity loop and a position loop. The feedback device for the position loop is normally an encoder or resolver. The control device (in this case a PLC) closes the position loop. The motor drive closes the velocity loop. A command is issued by the PLC. It is a velocity command. The drive receives the analog command (typically -10 to +10 V). The summing junction compares the command and the feedback from a tachometer on the motor and generates an error command to the motor. The velocity is continually monitored and adjusted by the drive. The position is monitored and controlled by the PLC. The minus sign in the summing junction shows that the negative-feedback principle is used.

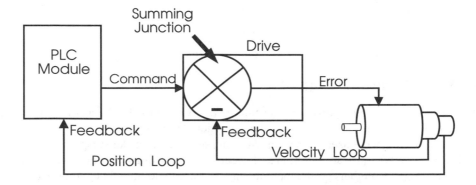

Figure 6-12. Closed-loop servo system.

Allen Bradley has developed an icon-based programming system for motion programming. Icons are small graphics that represent different types of motions or functions. The programmer just chooses the icons in the order which he/she wants the program to run and the software writes the actual application. The programmer also specifies speeds, feeds, accelerations/decelerations, and so on.

Vision Modules

Vision is becoming more and more widely used in industry. Inspection costs industry billions of dollars each year. Most of the inspection occurs after the product is made. That means that the product is already scrap. Automated vision allows for in-process 10 percent inspection. It also allows automatic adjustment of the process while it is running. This means that the process can be corrected before it makes scrap. This can drastically improve the quality of parts and productivity of processes.

Vision began to enter industry in the early 1980s. There were a multitude of new vision companies. Industry immediately saw the benefit that was possible through vision. Unfortunately, industry tried to apply vision to the wrong applications. Instead of trying to apply vision to the many tasks that were appropriate and easy to implement industry decided that they needed vision for automated welding, and automated guidance, and other applications. There were too many vision companies and all were looking for sales. Salespeople all said that their product could do anything. The horror stories began. Programming easily cost more than the vision equipment. The systems were not very good when misapplied. Vision achieved a bad name. Industry backed off on planned expenditures for vision. Many vision companies failed. In the process, however, the surviving vision companies retrenched. They chose niches of applications that were appropriate for their product. Industry learned from its early vision failures, too. Industry has now begun to choose applications more appropriately.

Vision systems can inspect at rates of thousands of parts per minute, at extremely high rates of accuracy. They are particularly suited to tasks at which people are not efficient. People are not very good at inspecting when the tasks are very quick, tedious, or involve rapid identification and/ or measurement. Vision systems do not get bored. For example, pharmaceutical manufacturers are required to inspect the date and lot code on every package they manufacture. There are no people, no matter how dedicated, that can inspect the high volumes that are required. A vision system is very well suited to this application.

The cost of vision is decreasing. This is due mainly to the fact that vision manufacturers have improved their software. They have made them much easier to program and apply. Thus huge reductions have occurred in application costs.

PLC manufacturers have seen the increased interest and now offer vision modules for their PLCs. Allen Bradley, for example, offers a full range of vision equipment: stand-alone and PLC vision modules.

Allen Bradley calls their product a Vision Input Module (VIM). The VIM, which can be plugged into several of the Allen Bradley PLC families, can inspect pieces at up to about 1800 pieces per minute. The VIM can be used for part presence sensing, measurement, alignment, and other inspection tasks.

The VIM module does not utilize a programming language. The programmer just "teaches" the system. The programmer places a good part in front of the camera. The camera provides a picture of the image to the video monitor. There are symbols (called icons) across the bottom of the screen. Each icon represents a different type of inspection task. The programmer just chooses the icons. Once the programmer has chosen icons to match the steps in the inspection application, the vision system is ready.

Omron also offers vision modules and stand-alone systems for visual inspection.

A vision system typically consists of the following components: one or more cameras, lighting, a video monitor, and the vision processor (see Figure 6-13). The lighting can often be the most difficult part of the application. The story is told of one vision company that designed and installed an application. The application worked perfectly. It continued to work perfectly for several months. Then, all of a sudden, it seemed to act up every day. The company contacted the vision company to come and fix the system. The vision technician struggled for several days and could not find any problem with the system, although every day it seemed to get erratic in the middle of the afternoon. Finally, the technician happened to glance at the ceiling. He noticed that there were some skylights that the sun was shining through. The company covered the windows and the system worked perfectly again. The time of year was just right for the sun to shine on the application in the middle of the afternoon. This story illustrates that lighting is a crucial part of every application. Many systems today will compensate for normal variations in ambient lighting.

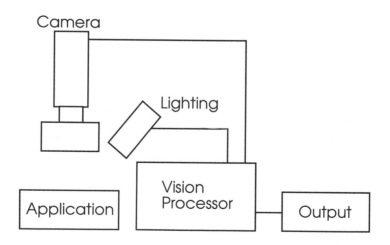

Figure 6-13. Block diagram of a typical vision system.

Bar-Code Modules

Automatic identification is becoming more widely used in industry. There are several types of automatic identification: vision, barcoding, and radio frequency identification (RFID). Barcoding is a rapidly growing technology. We see it in shopping stores, food stores, industry, and so on. It has become very commonplace.

Bar-code modules are available for PLCs. A typical application might involve a bar-code module reading the bar code on boxes as they move along a conveyor line. The PLC is then used to divert the boxes to the appropriate product lines. They are also used extensively to monitor inventories in factories and businesses. A factory is able to have real-time inventory counts. Imagine the information the grocery store has from barcoding every product and price into the database. They can measure very accurately the effect of advertising vs. sales of specific products, for example. They also use the information to handle which products and quantities must be ordered. The bar-coding also saves the cost of marking each and every product. There are several standard barcoding systems. The grocery business has standardized on the universal product code (UPC) standard.

PID Modules

Proportional, integral, and derivative (PID) control is used to control processes. It can be used to control physical variables such as temperature, pressure, concentration, and moisture content (see Figure 6-14). It is widely used in industrial control to achieve accurate control under a wide variety of process conditions. Although PID often seems complex, it need not be. PID is essentially an equation that the controller uses to evaluate the controlled variable. The controlled variable (temperature, for example) is measured and feedback is generated. The control device receives this feedback. The control device compares the feedback to the setpoint and generates an error signal. The error is examined in three ways. It is looked at with proportional, integral, and derivative methodology. The controller then issues a command (output signal) to correct for any measured error.

Proportional Control

The proportional part of PID looks at the magnitude of the error. Proportional control reacts proportionally. A large error receives a large response. (In the case of a large temperature error, the fuel valve would be opened a great deal.) A small error receives a small response.

Imagine a furnace that can be heated to 1500 degrees. There is a portion of this range of temperature where the response of the system will be proportional. For example, let's assume that between 1000 and 1500 degrees the system adjusts the valve opening in proportion to the error. Below 1000 degrees the valve is open 100 percent. Above 1500 degrees the valve is 0 percent open. The proportional band in this case is 500 degrees. Proportional band is normally given as a percentage. The percentage is calculated by dividing the proportional band in degrees by the full controller range and multiplying by 100. The full controller range is simply the range of temperatures that the furnace can control. In this case lets assume that the full controller range is 1440 (1500-60). Are you wondering why 60 was subtracted? The furnace cannot control below room temperature, so that is excluded in our calculation. In this case the proportional band in degrees is equal to (500/1440)X100 or 34.7 percent. The technician can adjust the width of the proportional band to make the system more or less responsive to an error. The narrower the proportional band, the greater the response to a given error.

Proportional control does not solve all of our problems. The first problem is that proportional control cannot correct for very small errors (these errors are called offset). The second problem with proportional is that it cannot adjust its output based on the rate of change in the measured variable. For example, if you are driving your car on the highway at 55 miles per hour and the car in front of you stops, you may respond with 50 percent brake pressure. This is a proportional response. The car ahead is stopping, so you respond with 50 percent brake. This will not work if the car stops too quickly. We, of course, would give much more brake pressure when the rate of stopping of the car in front is greater. In other words, we thought we applied enough brake pressure but the distance between our cars keeps getting smaller. We naturally respond to rate of error change. Proportional control systems do not. Proportional systems only respond to the magnitude of the error, not to its rate of change.

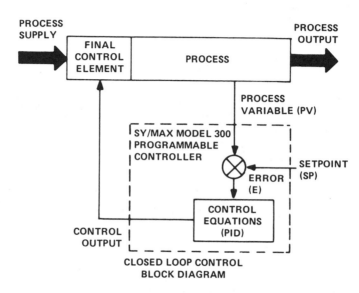

Figure 6-14. Block diagram of a process that is controlled by a PID controller. Imagine that the system is a furnace that controls temperature. The final control element would be the fuel valve. The process variable (feedback) is temperature. The setpoint is the temperature that the operator set. The feedback is compared to the setpoint and an error is generated. PID equations are then used to adjust the error to a suitable response (output). The controller in this case is a PLC. Courtesy of Square D.

Integral Control

The integral portion attempts to correct for the small error (offset) that proportional cannot. Integral looks at the error over time. It increases the importance of even a small error over time. Integral is error multiplied by the time the error has persisted. A small error at time zero has zero importance. A small error at time 10 has an importance of 10 X error. In this way integral increases the response of the system to a given error over time until it is corrected. Integral can also be adjusted. The integral adjustment is called <u>reset rate</u>. Reset rate is a time factor. The shorter the reset rate the quicker the correction of an error. In hardware-based systems the adjustment will be done by a potentiometer. The potentiometer essentially adjusts the time constant of an <u>RC</u> circuit. Many systems now employ software based control in which the technician changes a software parameter value for each factor. Too short a reset rate can cause erratic performance.

Chapter 6: Input/Output Modules and Wiring

Derivative Control

The derivative portion of control attempts to look at the rate of change in the error (see Figure 6-15). Derivative will cause a greater system response to a rapid rate of change that to a small rate of change. Think of it this way: if a system's error continues to increase, the control device must not be responding with enough correction. Derivative senses this rate of change in the error and causes a greater response. Derivative is also adjusted as a time factor. Derivative is also called rate time. Too much derivative can also cause erratic control.

$$Co = K(E + 1/Ti \int_{o}^{t} Edt + KD(E-E(n-1))/dt) + bias$$

Co	The output of the equation
K	Overall controller gain
1/Ti	Reset gain (integral factor)
KD	Rate gain (derivative factor)
dt	Time between process samples
bias	Output bias
E	System error. This is equal to the setpoint minus the measured value.
E(n-1)	This is the error of the last sample.

Figure 6-15. PID equation.

Most larger PLCs offer PID capability. Some offer the capability through special PID modules. Others offer PID ladder logic that can be used with standard modules.

Fuzzy Logic Modules

Fuzzy logic promises to become a much larger part of industrial control. (Some would say that most of us have always used "fuzzy logic.") Fuzzy logic is a very new technology in electronic control. It is being widely used in consumer electronics.

Much of industrial decision making is based on binary logic: for example, the temperature is either right or wrong, the device is either in position or out of position. The motor is on or off. The heating coil is on or off. PID control is used to control variables more closely. PID still relies on fixed formulas. The operator or programmer can adjust the gains of the PID, but then the controller still works with a fixed formula.

Fuzzy logic attempts to make decisions more like those of human beings. People make decisions based on multiple inputs. They do not make hard-and-fast decisions based on equations. Humans are able to vary the decision based on relative importance of inputs and other influences. They also understand concepts such as warm, cool, lukewarm, and so on.

People tend to make decisions in a "fuzzy" way. They can take in many inputs and decide on the relative importance of each based on present conditions. Imagine carrying a hot paper cup of coffee that was filled too full. A person can change how it is carried to adjust for speed of travel, going around a corner, going up/down stairs, making a quick stop, and so on. We adapt readily to new situations. We do this seemingly without thinking. We are, however, adjusting our acceleration/deceleration, speed, inclination of the cup, and so on. Fuzzy logic attempts to give hardware this capability.

One example of fuzzy logic control is the video camera. Some video recorders are now equipped with electronic circuitry to eliminate the unwanted movement of the camera that makes the video "jumpy." This is a relatively tough problem to solve. A person cannot hold a camera still. If the camera could compensate for the unwanted movement the jumpiness could be eliminated. How does the camera differentiate between desired movement and undesired movement? Fuzzy logic is used.

The process of fuzzy logic can be divided into two functions: inference and conclusion. Even a very complex decision is really just a set of simpler decisions. If each of the simpler questions is answered, the overall decision can be made based on the sum of the conclusions. In fuzzy logic the user sets up some parallel decisions.

For the fuzzy logic controller these simple decisions are called <u>rules</u>. Each rule is analyzed and a decision is made. The conclusion from each rule is then summed to achieve a logical sum. This logical sum is then "defuzzed" mathematically. This result is then used to calculate a value that the control device can use for output. Omron offers fuzzy logic modules for PLCs. They can be programmed to control a wide variety of industrial applications. Fuzzy logic is covered in greater detail in Chapter 7.

Radio-Frequency Modules

The use of radio-frequency (RF) modules is rapidly expanding. They are used for identification, data collection, production control, product tracking, and more. They eliminate some of the problems associated with bar codes. Bar codes must be kept clean. The air must be relatively clean, too. Radio-frequency modules are unaffected by dirt, oil, or any other common contaminant. They are also unaffected by plant noise. Manufacturing environments are filled with electronic noise. RF is unaffected by it. RF has achieved fairly wide use in automatic-guided vehicles (AGVs). They are used to send instructions to the vehicles as they "roam" factories. Present AGVs typically follow wires in the floor. Newer systems do not require a wire for guidance.

RF is also widely used in retail stores. The small plastic tags on expensive items such as leather jackets and tennis rackets are RF tags. They are single bit tags. They only need to be sensed by the RF module to set off an alarm if you take the item out of the store with the tag.

Another advantage of RF is that they feature bidirectional communication. They can be used to read and write. Imagine an RF tag on a car as it moves through production. The tag could easily contain all of the specifications for the car, including the options that the customer had ordered. As it moves through production, each station reads the RF tag so that the correct options are installed (see Figure 6-16).

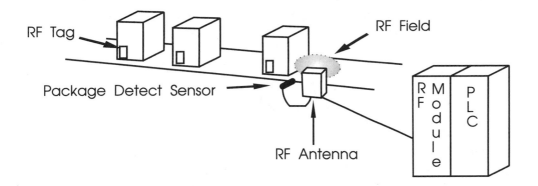

Figure 6-16. Typical RF application. As the pallets travel down the conveyor, they enter the RF field. A product sensor detects the product. The sensor is connected to the antenna. The antenna is triggered by the product sensor to read/write to/from the tag. The use of a product sensor helps avoid "false reads."

The same tag can be written to. So as the car moves along, quality tests are run and the results are written to the tag. When the car reaches the end of the line a complete production history is on the tag. The tags are capable of large amounts of storage. The typical tags available for PLCs feature 100 to 2000 bytes of memory.

There are two types of tags, active and passive. Active tags contain a battery; passive tags do not. As a passive tag passes the transmitter/receiver, the field that the transmitter radiates is used to power the tag. The only disadvantage to the passive tag is that the sensing distance is much less than the active tag. The batteries in active tags can last for several years. Active tags can work at much greater distances.

Radio frequency is catching on fairly rapidly. It is an "invisible" technology. With a bar code we can see the code, whereas we cannot see RF systems communicate. This has in some ways hampered the acceptance of RF technology. People do not understand things that they cannot see. They are also somewhat fearful or distrustful of invisible technology.

Operator Input/Output Devices

As systems become more integrated and automated, they become more complex. Operator information becomes crucial. There are many devices available for this information interchange.

Operator Terminals

Many PLC makers now offer their own operator terminals. They can be very simple to very complex. The simpler ones are able to display a short message. The more complex models are able to display graphics and text in color while taking operator input from touchscreens, bar codes, keyboards, and so on. These display devices can cost from a couple of hundred dollars up to several thousand.

If we remember that the PLC has most of the valuable information about the processes it controls in its memory, we can see that the operator terminal can be a window into the memory of the PLC.

The greatest advances have been in the ease of use. Many have software available that runs on an IBM personal computer. The software essentially writes the application for the user. The user draws the screens and decides which variables from the PLC should be displayed. The user also decides what input is needed from the operator. When the screens are designed, they are downloaded to the display terminal.

These smart terminals can store hundreds of pages of displays in their memory. The PLC simply sends a message that tells the terminal which page and information to display. This helps reduce the load on the PLC. The memory of the display is used to hold the display data. The PLC only requests the correct display, and the terminal displays it.

The PLC only needs to update the variables that may appear on the screen. The typical display would include some graphics showing a portion of the process, variables showing times or counts, and any other information that might aid an operator in the operation or maintenance of a system.

Speech Modules

One of the more interesting and unique operator information systems are speech modules. These modules are used to output messages to operators (see Figure 6-17).

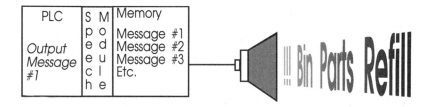

Figure 6-17. How a message is converted and stored in the memory of the speech module. The human voice is an analog signal. The analog acoustic signal is converted to an electrical analog signal by a microphone. The speech module uses an analog to digital converter to change the signal to a digitized file. The digitized message is given an address and stored in memory. The explanations about speech modules are based on the Omron Electronic's speech module.

Speech modules typically are used to digitize a human voice pronouncing the desired word, phrase, or sentence. The digitized sound is stored in the module's memory. Each word, phrase, or sentence is given a number. Ladder logic is used to output the appropriate message at the appropriate time. The sound from such modules is remarkably good.

There are two modes that can be utilized for output. Sentence mode would play a recorded sentence. Phrase mode would play selected words in the proper order to form a message. For example, assume that there are five machines in a cell. We might need a message to maintain the machine, refill the machine, and so on. Instead of storing a complete message for each case for each machine, phrases are combined to form a complete sentence (see Figure 6-18).

Message Address	Message
100	Maintenance required on Machine Number
101	Refill Parts Bin on Machine
102	Shutdown Machine Number
103	One
104	Two
105	Three
106	Four
107	Five
108	Thankyou

Figure 6-18. How the phrase mode of a speech module would be used. Individual messages would be combined to form a complete order. For example, if the message order was 102, 106 , the complete message would be "Shut down machine number four." After the operator shut down number four and acknowledged it to the PLC, the PLC might output message number 108, "Thank you." (Machines can be more polite than people.)

Special-purpose modules are available for just about any application. These modules can be used to greatly simplify the task of developing a system. As PLCs gain more capability, more and more special-purpose modules will become available. The emphasis will also shift to making the modules easier to use. The modules that will become more widely used are those that provide information to and from systems. The data these modules can provide can be used to improve the efficiencies of manufacturing processes.

Questions

1. What voltages are typically available for I/O modules?

2. If an input module was sensing an input from a load-powered sensor when it should not, what might be the possible problem, and what could you do about it?

3. What types of output devices are available in output modules? List at least three.

4. Explain the purpose of A/D modules and how they function.

5. Explain the purpose of D/A modules and how they function.

6. What type of module could be used to communicate with a computer?

7. If noise were a problem in an application, what types of changes might help alleviate the problem?

8. What are remote I/O modules?

9. Explain the term <u>resolution</u>.

10. With information becoming more important, which modules do you think will become more widely used? Explain your rationale.

11. What are the differences between RF and bar-code systems? Why might one be chosen over another?

12. Name the typical components in an RF system, and describe the fundamentals of its operation.

13. What is the difference between passive and active tags? Why might one be chosen over the other?

Chapter 7

Arithmetic Instructions

Arithmetic instructions are vital in the programming of systems. They can simplify the task of the programmer. In this chapter we cover compare , add, subtract, multiply, and divide instructions. Several brands of PLC instructions will be covered.

Objectives

Upon completion of this chapter, the student will be able to:

Describe typical uses for arithmetic instructions.

Explain the use of compare instructions.

Explain the use of arithmetic instructions such as add, subtract, multiply, and divide.

Write ladder logic programs involving arithmetic instructions.

Explain the similarities and differences between the various brands of PLC instructions.

Introduction

There are many times that contacts, coils, timers, and counters fall short of what the programmer needs. There are many applications that require some mathematical computation. For example, imagine a furnace application that requires the furnace to be between 250 and 255 degrees (see Figure 7-1). If the temperature variable is between 250 and 255 degrees, we might turn off the heater coil. If the temperature is below 250 we would turn on the heater coil. If the temperature is between 250 and 255 degrees, we turn on a green indicator lamp. If the temperature falls below 240 degrees, we might sound an alarm. (Note: Industrial temperature control is normally more complex than this. Complex process control is covered in Chapter 8.)

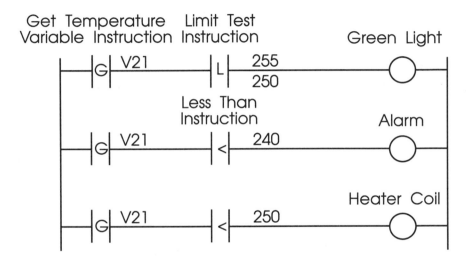

Figure 7-1. How comparison instructions could be used to program a simple application. In this application the heater coil will be on if the temperature is below 250 degrees. An alarm would sound if the temperature drops below 240 degrees. A green indicator lamp is on if the temperature is between 250 and 255 degrees.

This simple application required the use of many relational operators (arithmetic comparisons). The application involved tests of equality, less than, and greater than. The use of arithmetic statements makes this a very easy application to write. Many of the small PLCs do not have arithmetic instructions available. All of the larger PLCs offer a wide variety of arithmetic instructions.

There are also many cases when numbers need to be manipulated. They need to be added, subtracted, multiplied, or divided. There are PLC instructions to handle all of these computations. For example, we may have a system to control the temperature of a furnace. In addition to controlling the temperature we would like to convert the Celsius temperature to degrees Fahrenheit for display to the operator. If the temperature was to rise too high, an alarm would be triggered. Arithmetic instructions could do this very easily. An example of this is shown later in this chapter.

Allen Bradley Arithmetic Instructions

Allen Bradley controllers offer many arithmetic instructions. The standard instructions include addition, subtraction, multiplication, and division. There are many more instructions available.

Addition -(+)-

The addition instruction adds two values that are "fetched" by two get instructions in the rung. A get instruction is used to "get" a word (a word in this case is a number that is 16 bits long) from a memory location. A get instruction is used in a rung to "get" data from memory. A get instruction "gets" 16 bits of data from one location in memory. Although the instruction looks like a contact, it does not determine a rung's condition. Study Figure 7-2. If contact 11501 is true, the whole rung is true. The get instructions merely tell the processor to fetch two numbers from memory. Get instructions can be programmed at the beginning of the rung or other conditions can precede it. The number that is fetched by a get can then be used in some way by another instruction. In this case it is used to get two values to be added together. The decimal number fetched by the get will be displayed below the get instruction. The result of the add is stored in the word address of the add instruction. In this case it will be stored in address 010. The add instruction is programmed in the output position of the rung.

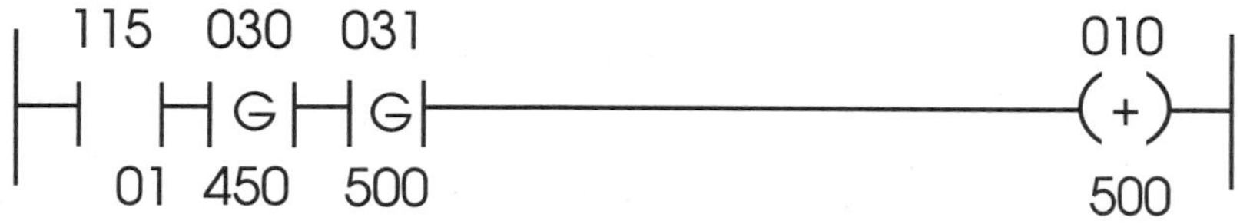

Figure 7-2. Allen Bradley add instruction. If the first contact (11501) is true, the add instruction will be performed. In this example the numbers (450 and 500) are "fetched" from addresses 030 and 031. They are then added. The result is stored in address 010.

Subtraction -(-)-

The subtraction instruction subtracts the two values that are "fetched" by two get instructions (see Figure 7-3). The second word value is subtracted from the first word value. The result is stored in the word address given by the subtraction instruction. Only positive values should be used.

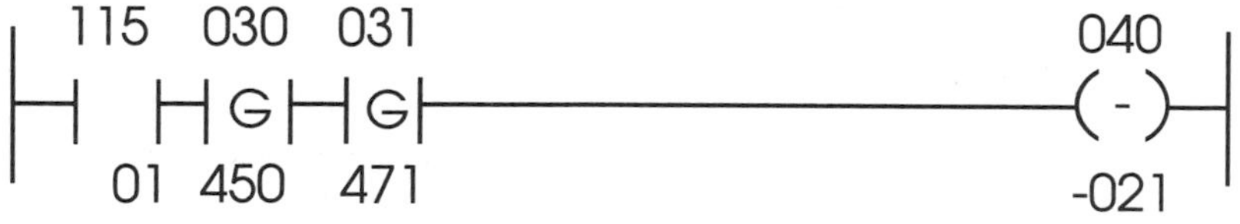

Figure 7-3. Allen Bradley subtraction instruction. If the first contact (11501) is true, the instruction will be performed. In this example the numbers (450 and 471) are "fetched" from addresses 030 and 031. They are then subtracted. The result is stored in address 010.

Multiplication -(X)-(X)-

The multiplication instruction is used to multiply two values that are "fetched" by get instructions in the rung. The result is stored in the two addresses specified by the multiplication instruction. The multiplication instruction is programmed in the output position of the rung. The addresses to which the result will be stored should be consecutive. Note that the result is stored in the format XXX,XXX (see Figure 7-4).

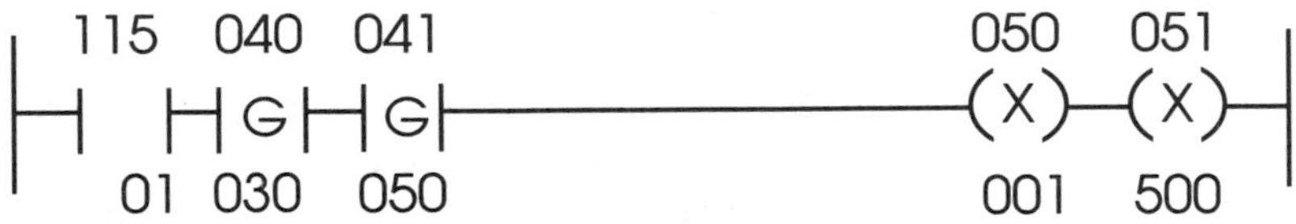

Figure 7-4. Allen Bradley multiply instruction. If the first contact (11501) is true, the multiplication instruction will be performed. In this example the numbers (30 and 50) are "fetched" (by two get instructions) from addresses 040 and 041. They are then multiplied. The result is stored in addresses 050 and 051.

Division

The division instruction is used to divide two values. The two values are "fetched" by two get instructions in the rung. The result of the division is stored in two word addresses that are specified by the division instruction. The addresses used for storing the result should be consecutive. Study Figure 7-5 to see how the result is stored. The whole portion of the result is placed in the first memory address (50). The decimal portion of the result is placed in the second address (51).

Figure 7-5. Allen Bradley division instruction. If the first contact (11501) is true, the division instruction will be performed. In this example the numbers (40 and 10) are "fetched" from addresses 040 and 041. Forty is then divided by 10. The result is stored in addresses 050 and 051. Note that the division is shown to three-decimal-place precision (XXX.XXX).

Allen Bradley Compare Instructions

The compare instruction compares the data in a memory location with data from another location. The result of the compare determines the rung condition. The get instruction is used to get the first data value to be compared.

Equality Comparison -[=]-

Two values can be compared to check if they are equal. The result determines the rung condition. If true the rung is true and the output will be energized; if false, the rung is false and the output is deenergized (see Figure 7-6).

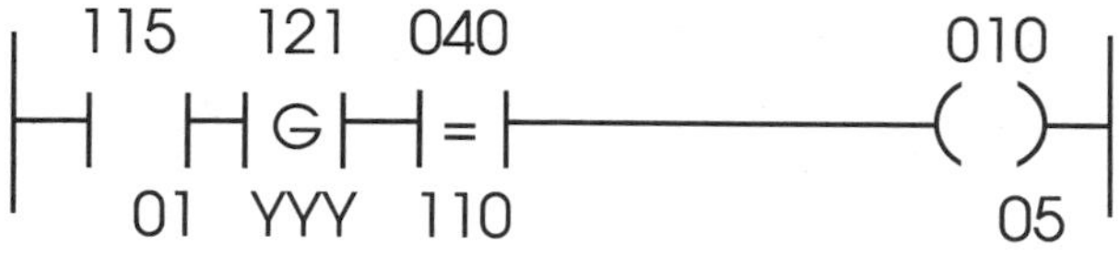

```
  115   121  040                  010
 ┤ ├ ┤ ├G┤ ├=┤ ├──────────────( )──┤
   01   YYY  110                   05
```

Figure 7-6. Use of a comparison instruction to check for equality.

Less Than Instruction -|<|-

The less than instruction is used to compare two values (see Figure 7-7). If the get value is less than the reference value stored in the less than instruction, the instruction is true and the output is energized.

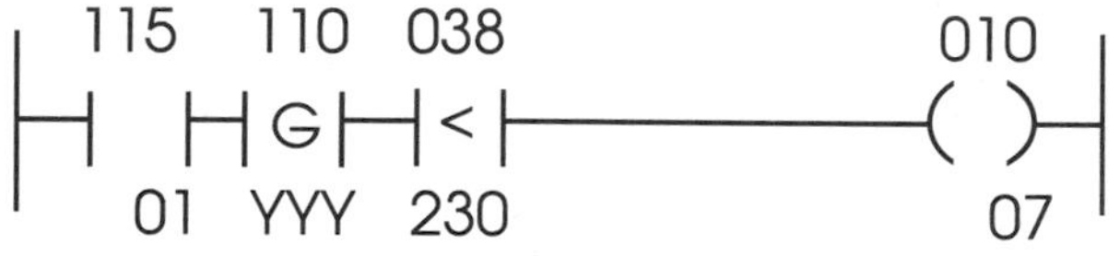

```
  115   110  038                  010
 ┤ ├ ┤ ├G┤ ├<┤ ├──────────────( )──┤
   01   YYY  230                   07
```

Figure 7-7. Less than comparison. The get instruction is used to get the value that is held in register 110. If the number held in 110 is less than 230, output 01007 will be energized. These instructions will be performed only if input 11501 is true.

Limit Test -(L)-

The limit test instruction is used to determine if the value of a byte is between two reference byte values (see Figure 7-8). In this case the byte fetched from 110 will be checked to see if it is less than 150 AND greater that 100. If the get value is between the values of the reference values, the instruction is true and the output energizes. Note that this instruction uses only one byte. Study Figure 7-8. Instead of the "G" we used to get a word from memory, a "B" is used. The "B" fetches one byte from memory. The normal get instruction "gets" a word (two bytes) from memory. The get byte instruction must be used. The get byte instruction -|B|- gets one byte from one word of memory. The data are shown in octal. In fact, the limit test and the get byte instruction work only in octal.

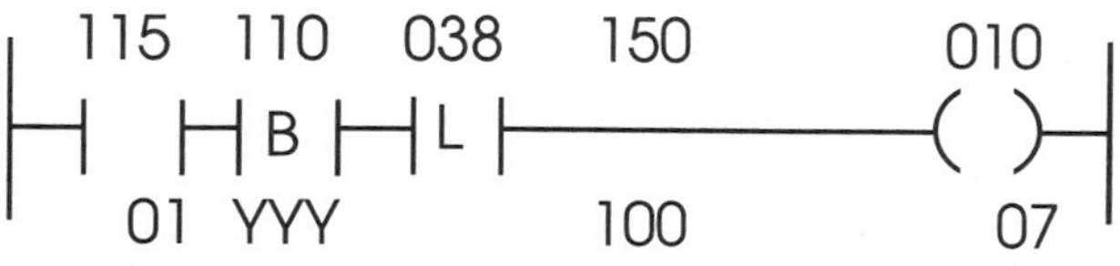

```
  115   110  038    150           010
 ┤ ├ ┤ ├B┤ ├L┤ ├──────────────( )──┤
   01   YYY         100            07
```

Figure 7-8. Use of a limit instruction. If input condition 11501 is true, a byte from address 110 will be "fetched." The value from 110 will then be compared to the limit values. If it is less than 150 and more than 100, output 01007 will be energized.

The programmer can use arithmetic instructions such as compares in series or in parallel in ladder logic. The programmer is then able to test for multiple conditions. It is almost like constructing your own instruction to accomplish a new task. Study Figure 7-9. The programmer needs to determine if a number was less than OR equal to the number 235. There is no less-than or equal-to instruction, so the programmer created one. In this case a less-than instruction and an equal-to instruction were used in parallel. If either of them is true, the rung is true and the output will be energized.

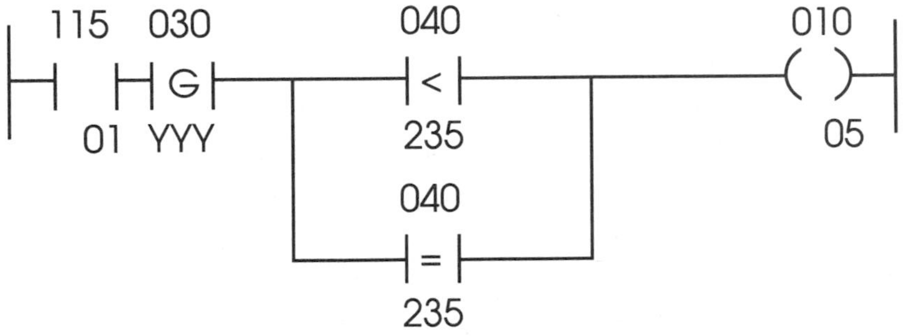

Figure 7-9. Use of comparison instructions in parallel. In this case if input 11501 is true, then the get instruction will "fetch" the number stored in address 30. If the value from 30 is less than 235, OR is equal to 235 output 01005 will be energized. These instructions can be programmed in series or parallel or in combination as long as the screen limits are not exceeded. This is a good way to test for multiple conditions.

The rungs shown in Figure 7-10 operate as follows:

Rung 1: The get instruction at address 200 is used to multiply the Celsius temperature 100 by 9. The result (900) is stored in address 203.

Rung 2: The get instruction at address 203 is used to divide 900 by 5. The result (180) is stored in address 205. Remember that had there been a decimal remainder, it would have been stored in address 206.

Rung 3: The get instruction at 207 is used to add 32 to the value 180, which is located at get address 205. The sum (212) is stored at address 210. The ladder so far has converted 100 degrees Celsius to 212 degrees Fahrenheit.

Rung 4: This rung gets the temperature from address 210. If the value is less than 190, the timer (T33) begins timing for 3 seconds.

Rung 5: After 3 seconds has elapsed, the output at address 01115 will energize. The output is a heating coil. The heater will bring the temperature back into the desired range.

Rung 6: The counter in this rung (counter 034) is used to count the number of times that the system temperature falls below 190 degrees. Input 03317 (timer 33, status bit 17) is used to increment the counter. Every time the timer energizes, bit 17 of the timer energizes.

Rung 7: When the temperature is equal to 212 degrees, a latching output (01116) energizes. The output could be an alarm, light, or any other device.

Rung 8: Rung 8 enables an operator to acknowledge the alarm condition. If an operator closes input switch 11014, output 01116 is unlatched and is deenergized.

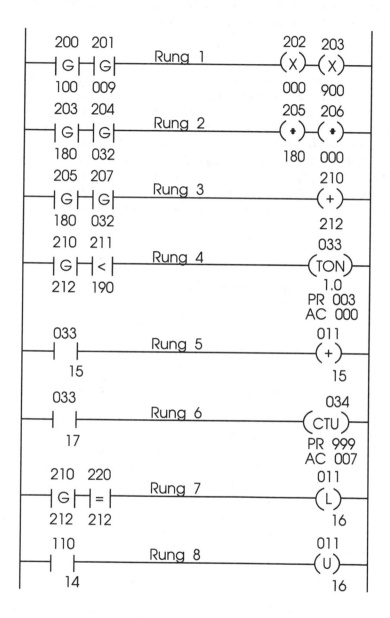

Figure 7-10. Ladder diagram used to control temperature. The logic first converts 100 degrees Celsius to 212 degrees Fahrenheit. If the temperature falls below 190 degrees Fahrenheit a heating coil is turned on. If the temperature reaches 212 an alarm is latched on until an operator acknowledges the alarm condition.

Gould Modicon Arithmetic Instructions

Gould Modicon uses a function block style for arithmetic instructions. There are many Arithmetic instructions available. Only a few are covered here.

Addition Instruction

The addition instruction is used to add two numbers and store the result to an address (see Figure 7-11). The addition instruction requires three entries from the programmer. The top entry is added to the middle entry and the result is stored in the register shown by the bottom entry. The top two entries can be an actual number or a 4XXX or a 3XXX register. If actual numbers are used, they can be between 0 and 999. If registers are used for the top two entries, the instruction gets the numbers from those two addresses and adds them. The result is stored in the register shown by the bottom entry. The programmer must also supply an input condition. The instruction is executed every time the rung is scanned. The use of a transitional input can assure that the instruction is executed only for transitions, instead of continuously if the input condition is true.

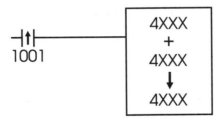

Figure 7-11. Gould Modicon addition instruction. The X's represent digits.

See Figure 7-12 for an example of the use of an addition instruction. Imagine that a PLC is controlling a bottling line. The manufacturer would like to count the number of bottles produced by the line. An addition instruction is used to keep track of the total number produced. Every time a case is full, a sensor is triggered as the case leaves the packing station. The output from the sensor (input 1001) is used as an input coil to trigger the addition instruction. The addition instruction then adds 24 to the total count. The total count is being held in register 4002. Note that one actual value was used and two addresses (both the same address). The same address was used for both so that a running total of bottles can be kept. (If input 1 becomes true, 24 is added to the current total from address 4002 and then the new total is put back in 4002.)

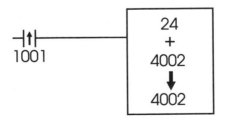

Figure 7-12. Gould Modicon addition instruction.

Subtraction Instruction

The subtraction instruction is very similar to the addition instruction. The subtraction instruction requires three values. The middle value is subtracted from the top value and the result is stored in the register shown in the bottom entry (see Figure 7-13). The subtraction instruction is much more versatile than the addition instruction. The subtraction instruction has three outputs available. The subtraction instruction can actually be used to compare the size of numbers. The top output is energized when the value of the top element is greater than the middle element. The middle value is energized when the top and middle elements are equal. The bottom output is energized when the top value is less than the middle value.

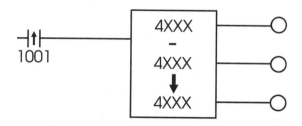

Figure 7-13. Gould subtraction instruction.

Multiplication Instruction

The multiplication instruction also uses a block format. Three entries are required. The top and middle entries are multiplied by each other and the result is placed in the register shown by the bottom entry. The top and middle entries can be an actual number or a register (see Figure 7-14). The third entry the programmer makes is the register where the answer is to be stored. The Gould Modicon uses two addresses to store the result in case the number gets large. For example, if the programmer designated register 4002 for the bottom entry, the PLC would store the result in registers 4002 and 4003. The programmer should keep this in mind when using registers. Do not use the same register for two different purposes unless there is a valid reason.

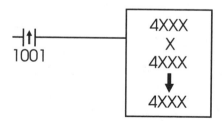

Figure 7-14. Gould multiplication instruction.

Division Instruction

The division instruction also uses a block format (see Figure 7-15). There are three entries. The top and middle elements can be actual numbers or registers. The third entry is where the answer will be stored. It must be a register. The division instruction divides the top number by the middle number. Note that a transitional input was used for the input to this instruction. This instruction uses two registers for the top entry. This means that if register 4002 was programmed for the top value, the number to be divided would be retrieved from registers 4002 and 4003. The output of the instruction is energized any time the result is too large for the storage register. The output could be used for error checking by the programmer.

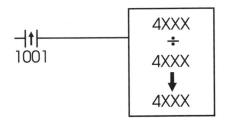

Figure 7-15. Gould divide instruction.

Omron Arithmetic Instructions

Omron offers a wide variety of arithmetic instructions for binary and BCD numbers. For all of these instructions the programmer provides numbers, or sources of numbers, and the address where the result will be stored. There are instructions available for binary and BCD numbers. A few of these are described in the next few pages.

Omron C200H Memory Areas

Omron divides their memory up into areas (see Figure 8-16). There are several types of areas, each with a specific use.

Omron Data Areas	
HR	Holding Relay Area
TR	Temporary Relay Area
AR	Auxiliary Relay Area
LR	Link Relay Area
TC	Timer/Counter Area
DM	Data Memory Area
IR	I/O and Internal Relay Area
SR	Special Relay Area
UM	Program Memory

Figure 7-16. This figure shows the memory areas for the Omron C200H series.

Holding Relay Area

The HR (holding relay) area is used to store and manipulate numbers. It retains the values even when modes are changed or during a power failure. The address range for the HR area is 0000 to 9915. This memory area can be used by the programmer to store numbers that need to be retained.

Temporary Relay Area

The TR (temporary relay) area is used for storing data at program branching points.

Auxiliary Relay Area

The AR (auxiliary relay) area is used for internal data storage and manipulating data. A portion of this data area is reserved for system functions. The AR bits that can be written to by the user range from 0700 to 2215.

Link Relay Area

The LR (link relay) area is used for communications to other processors. If it is not needed for communications, it can be used for internal data storage and data manipulation. The address range for the LR area is 0000 to 6315.

Timer/Counter Area

The TC (timer/counter) area is used to store timer and counter data. The timer counter area ranges from 000 to 511. Note that a number can be used only once. The same number cannot be used for a timer and a counter. For example, if the programmer uses the numbers 0 to 5 for timers, they cannot be used for counters.

Data Memory Area

The DM (data memory) area is used for internal storage and manipulation of data. It must be accessed in 16-bit channel units. If a multiplication sign is used before the DM (*DM), it means that indirect addressing is being used. Data memory ranges from addresses 0000 to 1999. The user can only write to addresses 0000 to 0999.

I/O and Internal Relay Area

The IR area is used to store the status of inputs and outputs. Any of the bits that are not assigned to actual I/O can be used as work bits by the programmer. Channels 000 to 029 are allocated for I/O. The remaining addresses up to 24615 is for work area. The IR area is addressed in bit or channel units. The addresses are accessed in channels or channel/bit combinations. Channel/bit addresses are 5 bits long. The two least significant digits are the bit within the channel. For example, address 01007 would be channel 10, bit (or terminal) 7.

Special Relay Area

The SR (special relay) area can be used to monitor the PLC's operation. It can also be used to generate clock pulses and to signal errors. For example, bit 25400 is a 1-minute clock pulse. Bit 25500 is a 0.1-second clock pulse. These special bits can be used by the programmer in ladder logic. There are many bits provided to signal errors that could occur. The programmer can use these in ladder logic to indicate problems.

ProgramMemory

The UM area is where the users program instructions are stored. Memory is available in various sizes (RAM and ROM).

These data areas are used simply by using the prefix (such as HR) followed by the appropriate address within that data area.

Binary Addition

To program a binary add instruction the user chooses the ADB instruction (function 50). The programmer enters the function number to specify which instruction will be used. The programmer then must supply the Au (augend) value, the Ad (Addend) value, and the R (result channel). If a "@" is used in front of the ADB (@ADB), the instruction is transitional. It will be active only when there is a transition on the input to it. The @ can be used with most of the instructions.

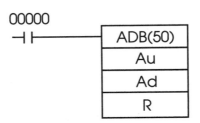

00000

ADB(50)
Au
Ad
R

Figure 7-17. Binary add instruction.

Figure 7-18 shows the data types that can be used in an add instruction. The "#" means that actual data are being entered. A hex number would be entered.

Au and Ad	R
IR, SR, HR, AR, LR, TC, DM, *DM, #	IR, HR, AR, LR, DM, *DM

Figure 7-18. Data types that can be used with a binary add instruction.

The use of the ADB instruction in a ladder diagram is shown in Figure 7-19. This figure shows that a ADB (function 41) is used. If input 00000 comes true, then the CLC (clear carry) instruction clears the carry bit. Input 00000 also causes the ADB instruction to execute. In this case the binary data from channel 1 will be added to the hex number #F8C5. The result will be stored in holding relay 21. The clear carry should be used to ensure that the addition is correct. The clear carry instruction should be used with any addition, subtraction, or shift instruction to ensure the correct result. Note that it can be programmed using the same input conditions as the actual add or subtract instruction. Figure 7-20 shows how the add instruction works. The Au and Ad are added and the result is placed into the R address. In this case the binary data for Au comes from actual input channel 1. These data are added to the hex number F8C5. The result is FA4C. The result is stored in HR21 (holding relay 21).

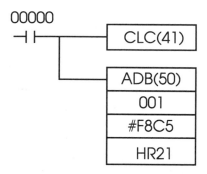

Figure 7-19. Use of the ADB (binary addition) instruction. Note the use of the clear carry instruction.

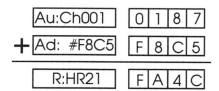

Figure 7-20. Results of a binary add instruction. The result is stored in address HR21.

Binary Subtraction

A binary subtraction instruction is programmed just like an add instruction. The instructions mnemonic is SBB (see Figure 7-21).

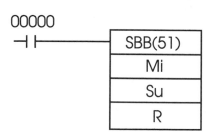

Figure 7-21. This figure shows a binary subtract instruction.

If it is entered as @SBB it will only be active when there is a transition on its input condition. The programmer enters a Mi (minuend), a Su (subtrahend), and a R address where the result will be stored. The same data types may be used for numbers to be subtracted as with the addition instruction. See Figure 7-18 for the possible types. The function code for this instruction is a 51. A BCD subtraction instruction is also available.

The Su (subtrahend) will be subtracted from the Mi (minuend). The result is stored in the area specified by the R value. The same types of areas can be used for each of these as was used for the addition instruction. Remember that a clear carry [CLC(41)] should be used to ensure that the correct answer is obtained.

Multiply Instructions

The binary multiply instruction (MLB) can be used to multiply two numbers. The function number for this instruction is 52 (see Figure 7-22). The programmer supplies a Md (multiplicand) and a Mr (multiplier) and the result is stored in the area specified by R. The R specifies the first of two addresses that will be used to store the result of the multiplication. This instruction multiplies two 16-bit numbers together and stores a 32-bit answer at channel R and R+1. The same data types can be used for the numbers to be multiplied as for the addition instruction. See Figure 7-18 for a list of the possible data types. A BCD multiply instruction is also available.

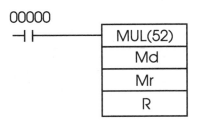

Figure 7-22. Binary multiply instruction.

Binary Division

The binary division instruction [DVB (53)] is programmed like the multiplication instruction. The programmer supplies a Dd (dividend), a Dr (divisor), and a R address. The R is the beginning address of the result. The quotient will be stored in channel R, and the remainder of the division problem will be stored in channel R+1. The same data types can be used as for the addition instruction. (See Figure 7-18 for the possible data types.) See Figure 7-23 for an example of the instruction.

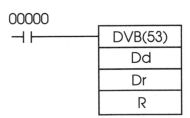

Figure 7-23. Binary divide instruction.

Square D Arithmetic Instructions

One of the simplest math instructions available is the "let" instruction. The let instruction can be used to assign a value to a variable: for example, let S46 = 100. It can also be used to perform arithmetic instructions such as addition, subtraction, multiplication, division, and square root (see Figures 7-24 to 7-27). The larger Square D models also perform sine, cosine, log, and absolute value operations.

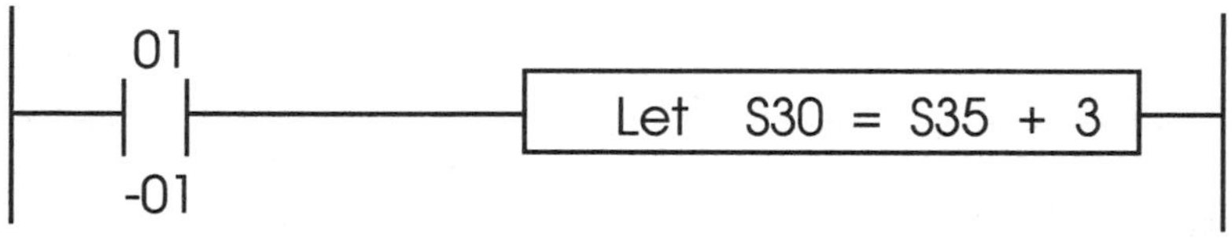

Figure 7-24. Let instruction used to add 3 to the number stored in register S35. The result is then stored in register S30. Note that the number subtracted could also have been a value stored by a register. The number held in register S35 does not change.

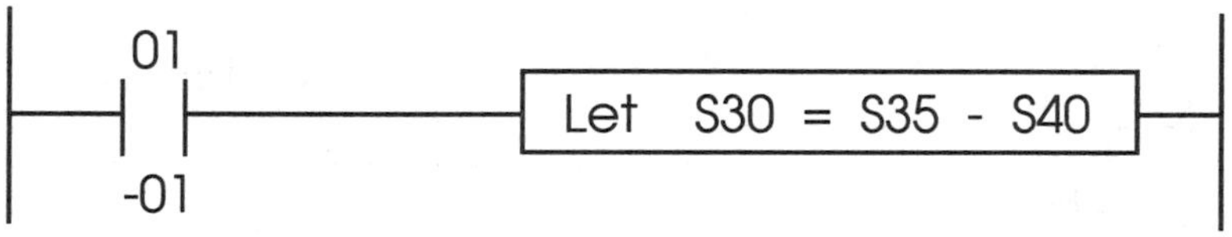

Figure 7-25. Let instruction used to subtract two numbers. The number in register S40 is subtracted from the value held in S35. The result is then stored in register S30. _Note_: The values on the right of the "equal" sign could have been actual numbers instead of registers that hold a number. The number held in S35 does not change.

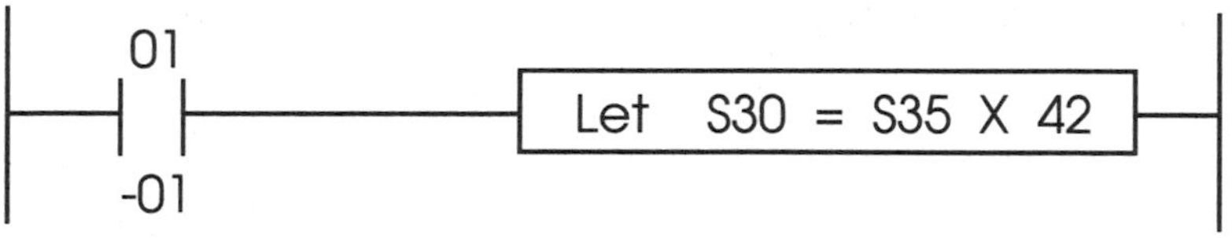

Figure 7-26. Use of a let instruction to multiply two numbers. In this case the number held by register S35 is multiplied by 42. The result is stored in register S30. Note that the number held in register S35 does not change.

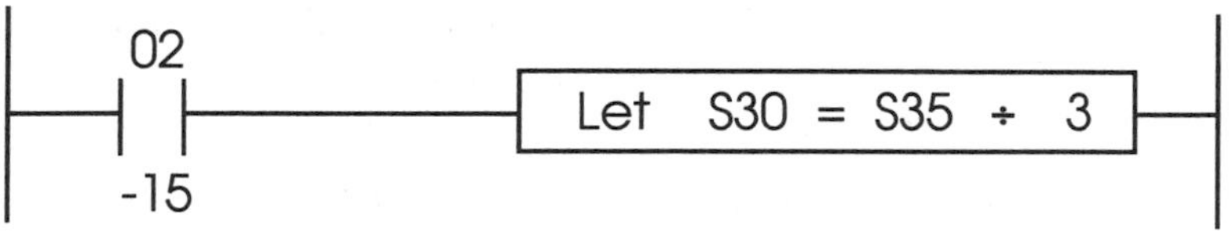

Figure 7-27. Use of a let instruction to divide two numbers. In this case the number held by register S35 is divided by 3. The result is stored in register S30. Note that the number held in register S35 does not change.

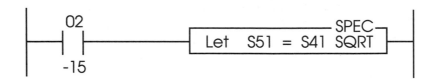

Figure 7-28. Let instruction used to find the square root of the number held in register S41. The result is then stored in register S51. The number held in register S41 does not change.

Let instructions can also be used to convert from one number system to another. The let instruction could be used to convert a BCD number to a binary number (see Figures 7-28,29, and 30). This could be used to convert inputs from BCD thumbwheel switches to a binary equivalent. Thumbwheel switches typically output BCD numbers. The BCD must then be converted to a binary for the PLC to understand it.

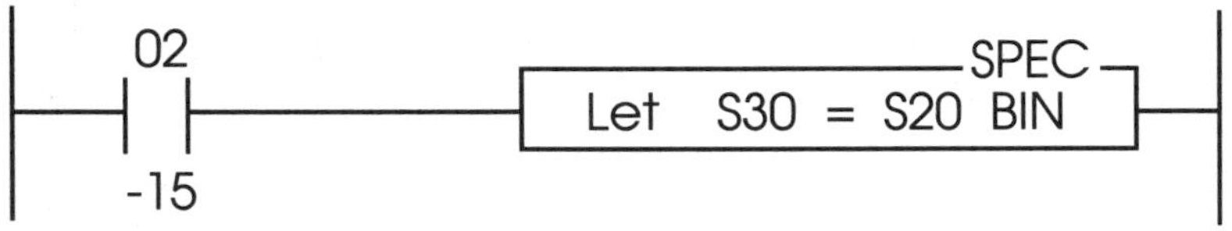

Figure 7-29. Use of a let instruction to convert the number held in register S20 to its binary equivalent. The result is stored in register S30. Note that the number held in register S20 does not change.

The let instruction can also be used to convert a binary number to a BCD number (see Figure 7-30). This is very useful when we need to convert a binary number to a decimal number for output to an operator. This could be used to output a BCD number that could drive a numeric display.

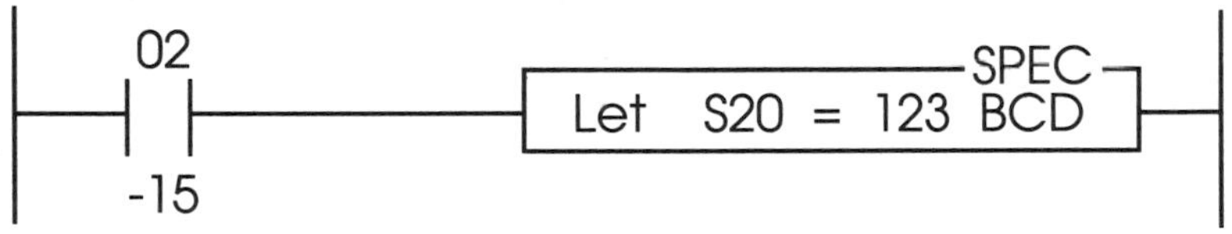

Figure 7-30. Use of a let instruction to convert a number to a BCD number. If input 0215 is true, 123 will be converted to BCD and stored in register S20. Note that instead of using an actual value the address of a register could have been used.

Figure 7-31. (Shown on the facing page.) How BCD thumbwheel switches could be used to input decimal numbers into a normal PLC input module. The thumbwheels output BCD. Each thumbwheel switch outputs 4 bits. Each bit corresponds to a bit in storage register S1. Note that S1 is the register that holds the status of real-world inputs 1-1 through 1-16. The BCD number is stored in register S1. A let instruction is used to convert the BCD to its binary equivalent and is then stored in register S2. The CRT would then display the proper decimal equivalent.

Chapter 7: Arithmetic Instructions

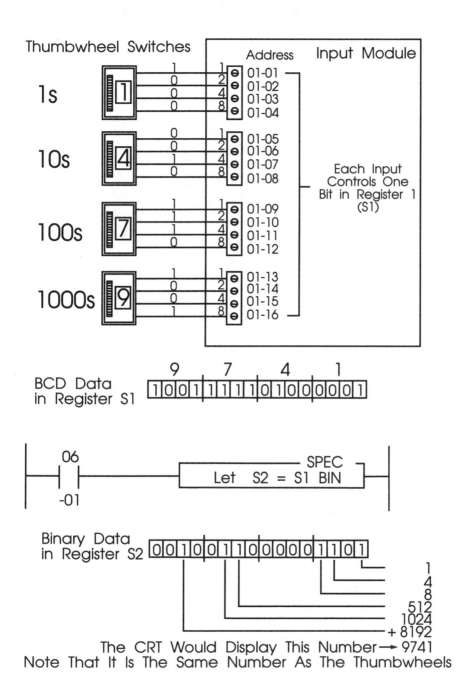

Thumbwheel Switches

1s

10s

100s

1000s

Address   Input Module

01-01
01-02
01-03
01-04

01-05
01-06
01-07
01-08

01-09
01-10
01-11
01-12

01-13
01-14
01-15
01-16

Each Input
Controls One
Bit in Register 1
(S1)

BCD Data
in Register S1

The CRT Would Display This Number → 9741
Note That It Is The Same Number As The Thumbwheels

Binary Data
in Register S2

1
4
8
512
1024
+ 8192

Square D Compare Instructions

The IF statement is used with Square D controllers to perform comparisons (see Figures 7-32 to 7-35). The IF instruction can be used to compare the number in a storage register to another storage register, a constant value, or the result of a math instruction. The IF can be used to test for equal, not equal, greater than or equal to, less than, or greater than.

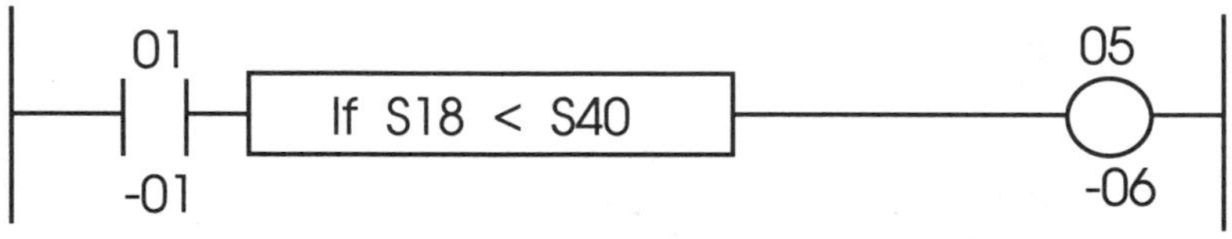

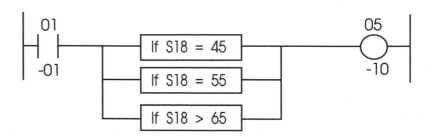

Figure 7-32. IF statement being used to test for a less-than condition. If the number held in storage register S18 is less than the number held in storage register S40, coil 5-06 will be energized.

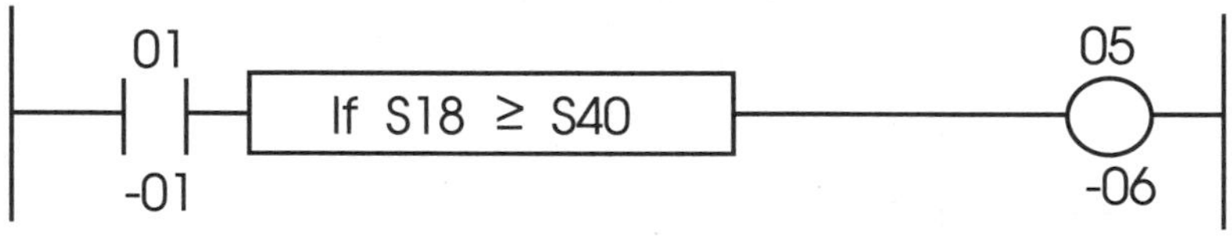

Figure 7-33. IF statement being used to test for a greater-than or equal-to condition. If the number held in storage register S18 is greater than or equal to the number held in storage register S40, coil 5-06 will be energized.

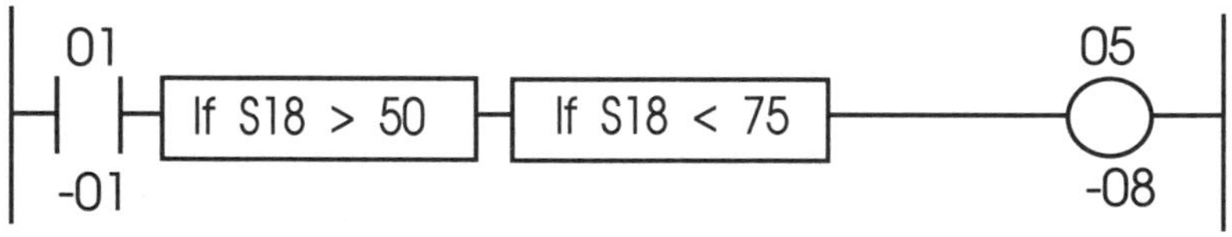

Figure 7-34. This figure shows the use of multiple IF statements in series (AND condition). If storage register S18 is greater than 50 AND less than 75, coil 5-08 will be energized.

Figure 7-35. Use of IF statements in parallel. If storage register S18 is equal to 45 OR equal to 55 OR greater than 65, coil 5-10 will be energized.

Let instructions can also be used for Boolean logic. Let instructions can perform ANDs, ORs, and XORs (Exclusive ORs).

Siemens Industrial Automation Arithmetic Instructions

There are many arithmetic instructions available for the 405 series. There are binary and BCD instructions. Many of these instructions make use of the accumulator. The accumulator is a memory location that is used to store numbers temporarily. Think of it as a scratch pad that can be used to hold a number. Numbers can be "loaded" into it, or the number in the accumulator can be stored to a different memory location. For example, the states of inputs can be loaded into the accumulator. They could then be manipulated and sent to actual outputs. There are many uses for the accumulator (see Figures 7-36 & 7-37).

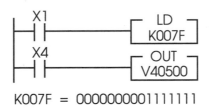

K007F = 0000000001111111

Figure 7-36 Use of a load instruction. If input X1 becomes true, the load instruction will load the accumulator with the hex number 007F. The K preceding the number stands for constant. If X4 becomes true, the number in the accumulator (007F) will be output to variable V40500. Variable V40500 is the word that controls the states of the first 16 outputs. The number below the ladder logic shows that the first 7 outputs would be energized.

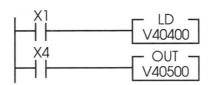

Figure 7-37. Use of a load instruction. If input X1 becomes true, the first 16 input states are loaded. Variable V40400 is a word that contains the states of the first 16 inputs. If input X4 becomes true, the number (word) held in the accumulator will be output to variable V40500. Variable V40500 is the word that controls the states of the first 16 outputs.

Omron arithmetic instructions utilize the accumulator. The use of an add instruction is shown in Figure 7-38. When input X5 becomes true the hex number 1324 will be added to whatever number is in the accumulator. The number that was in the accumulator is replaced by the result. If we wanted to add the 5 and 7 we would first use a load instruction to load 5 into the accumulator. We would then use an add instruction with a value of 7. The result of the instruction would be placed into the accumulator. We can use variables or constants with these instructions. In fact, the actual states of inputs, outputs, timers, and counters are all stored in variables. This means that we can utilize this data in instructions also.

Study Figures 7-39, 40, and 41 for examples of other arithmetic instructions.

Figure 7-38. Use of a add instruction. If input X5 becomes true, the hex number 1324 will be added to the value in the accumulator. The result is then stored back in the accumulator. The value that was in the accumulator is lost when the new value is stored there.

Figure 7-39. Use of a subtract instruction. If input X4 becomes true, the number in variable V1500 will be subtracted from the number in the accumulator. The result is stored in the accumulator.

Figure 7-40. Use of a multiply instruction. If input X3 becomes true, the number in variable V1530 will be multiplied by the number in the accumulator. The result is stored in the accumulator.

Figure 7-41. Use of a division instruction. If input X3 becomes true, the number in the accumulator will be divided by the number in variable V2320. The result is stored in the accumulator.

Questions

1. Explain some of the reasons why arithmetic instructions are used in ladder logic.

2. What are let instructions used for?

3. What are comparison instructions used for?

4. Why might a programmer use an instruction that would change a number to a different number system?

5. Write a rung of ladder logic that would compare two values to see if the first is greater than the second. Turn an output on if the statement is true.

6. Write a rung of logic that checks to see if one value is equal to a second value. Turn on an output if true.

7. Write a rung of logic that checks to see if a value is less than 20 or greater than 40. Turn on the output if the statement is true.

8. Write a rung of logic to check if a value is less than or equal to 99. Turn on an output if the statement is true.

9. Write a rung of logic to check if a value is less than 75 or greater than 100 or equal to 85. Turn on an output if the statement is true.

10. Write a ladder logic program that accomplishes the following. A production line produces items that are packaged 12 to a pack. Your boss asks you to modify the ladder diagram so that the number of items is counted and the number of packs is counted. There is a sensor that senses each item as it is produced. Use the sensor as an input to the instructions you will use to complete the task. (Hint: One way would be to use a counter and at least one arithmetic statement.)

Chapter 8

Advanced Programming

There are many instructions available that can make the programming systems easier. In this chapter we examine a few of them. Many of these instructions were designed to "look" like the mechanical control devices they replaced. The purpose was to make them easy to understand.

Objectives

Upon completion of this chapter, the student will be able to:

Explain a few of the advanced instructions that can make programming a complex application easier.

Explain such terms as <u>drum controller</u>, <u>interlocking</u>, <u>sequencers</u>, <u>stage programming</u>, and <u>step programming</u>.

Write simple sequencing programs using an appropriate advanced instruction.

Choose an appropriate advanced instruction for an application.

Sequential Control

Before PLCs there were many innovative ways to control machines. One of the earliest control methods for machines was punched cards. These date back to the earliest automated weaving machines. Punched cards controlled the weave. Until a relatively few years ago the main method of input to computers was punched cards.

Most manufacturing processes are very <u>sequential</u>, meaning that they process a series of steps, from one to the next. Imagine a bottling line. Bottles enter the line, are cleaned, filled, capped, inspected, and packed. This is a very sequential process. Many of our home appliances work sequentially. The home washer, dryer, dishwasher, breadmaker, and so on, are all examples of sequential control. Plastic injection molding, metal molding, packaging, and filling, are a few examples of industrial processes that are sequential.

Many of these machines were (and some still are) controlled by a device called a drum controller. A drum controller functions just like an old player piano. The player piano was controlled by a paper roll with holes punched in it. The holes represent the notes to be played. Their position across the roll indicates which note should be played. Their position around the roll indicates when they are to be played. A drum controller is the industrial equivalent.

The drum controller is a cylinder with holes around the perimeter with pegs placed in the holes (see Figure 8-1). There are switches that the pegs hit as the drum turns. The peg turns, closing the switch that it contacts and turning on the output to which it is connected. The speed of the drum is controlled by a motor. The motor speed can be controlled. Each step must, however, take the same amount of time. If an output must be on longer than one step, consecutive pegs must be installed.

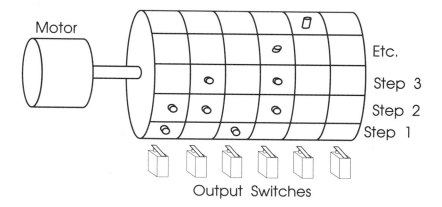

Figure 8-1. Drum controller. Note the pegs that activate switches as the drum is turned at slow speed by the motor. The sequence is changed by changing the peg locations. The speed is adjusted by changing the motor speed. Each step must have the same time interval. See Figure 8-2, which shows the output conditions for the steps.

The drum controller has several advantages. It is easy to understand, which makes it easy for a plant electrician to work with. It is easy to maintain. It is easy to program. The user makes a simple chart that shows which outputs are on in which steps (see Figure 8-2). The user then installs the pegs to match the chart.

Step	Input Pump	Heater	Add Cleaner	Sprayer	Output Pump	Blower
1	on		on			
2	on	on		on		
3		on		on		
4					on	
5						on

Figure 8-2. Output conditions for the drum shown in Figure 8-1.

Many of our home appliances are also controlled with drum technology. Instead of a cylinder they utilize a disk with traces (see Figure 8-3). Think of a washing machine. The user chooses the wash cycle by turning the setting dial to the proper position. The disk is then moved very slowly as a synchronous motor turns. Brushes make contact with traces at the proper times and turn on output devices such as pumps and motors.

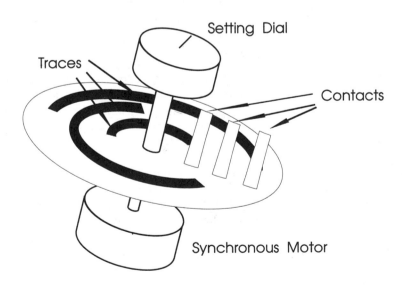

Figure 8-3. Disk type drum controller. It is commonly used in home appliances. Note that when you turn the dial you are actually turning the disk to its starting position.

Although there are many advantages to this type of control, there are also some major limitations. The time for individual steps cannot be controlled individually. The sequence is set and it does not matter if something goes wrong. The drum will continue to turn and turn devices on and off. In other words, it would be nice if the step would not occur until certain conditions are met. PLC instructions have been designed to accomplish the good traits of drum controllers and also to overcome the weaknesses.

Sequencer Instructions

Sequencer instructions can be used for processes that are cyclical in nature. Sequencer instructions can be used to monitor inputs to control the sequencing of the outputs. Sequencer instructions can make programming many applications a much easier task. The sequencer is very like the drum controller.

Allen Bradley Sequencer Instructions

Allen Bradley PLCs use function blocks to program sequencers. Three function block instructions are available: sequencer input, sequencer output, and sequencer load. Each sequencer block has up to 999 steps available. Each step can control up to 64 outputs. As you can see, some very complex applications can be programmed with these. Examine Figure 8-4. This figure shows a typical sequencer input instruction block. The data that the programmer inputs to the instruction are shown in Figure 8-5.

The counter address in the instruction indicates the address where the counter is stored. The current step indicates the present accumulated value of the counter. This instruction uses a counter to track which step it is in. The sequence length determines how many steps the sequencer has. (This is also the preset value of the counter.)

The words per step determines how many inputs are used. (Remember that one word is 16 bits, or inputs in this case.) The file input determines the address of the file. The mask input determines the address of the file that will be used as a mask. The last input shows which words will be used for the inputs. As this instruction is incremented through its steps, the inputs will be examined through the mask.

In actual use the input sequencer will be used as the input to an output sequencer instruction. For example, assume that the sequencer input instruction is in step one. If the input to the input instruction becomes true, the sequencer input instruction will increment to the next step. The inputs will be evaluated through the mask. (Remember that if the mask bit is a zero, the actual input condition for that bit is ignored.) If states of the inputs are correct, the input instruction will be true. The true output is used as an input to the output sequencer instruction. The output sequencer then increments to the next step. It outputs the new word of output conditions to the actual outputs. When the input sequencer instruction is enabled again, it will examine the states of the inputs that correspond to the new input step and continue the process.

The inputs that the programmer provides the output instruction are just like those of the sequencer input instruction. An example of a sequencer output instruction is shown in Figure 8-6. Figure 8-7 gives an explanation for each value that the programmer must input.

```
SEQUENCER INPUT
COUNTER ADDRESS: 0200
CURRENT STEP:         001
SEQ LENGTH:           006
WORDS PER STEP:         2
FILE:            0400- 0413
MASK:            0070- 0071

INPUT WORDS:
1:    0110    2:    0200
3:            4:
```

Figure 8-4. Sequencer input instruction.

COUNTER ADDRESS:	Address of the instruction.
CURRENT STEP:	Accumulated value of the counter. (Present step.)
SEQ. LENGTH:	The present value of the counter. (Number of steps)
WORDS PER STEP:	This is the width of the sequencer table.
FILE:	Starting address of the sequencer table.
MASK:	Address of the mask file.
INPUT WORDS:	These are the input words that are examined by the instruction.

Figure 8-5. Values required for a sequencer input instruction.

```
SEQUENCER OUTPUT
COUNTER ADDRESS: 0200
CURRENT STEP:         001
SEQ LENGTH:           006
WORDS PER STEP:         2
FILE:            0400- 0413
MASK:            0070- 0071

OUTPUT WORDS:
1:    0110    2:    0200
3:            4:
```

Figure 8-6. Format for a sequencer output block for an Allen Bradley PLC.

COUNTER ADDRESS: Address of the instruction.

CURRENT STEP: Accumulated value of the counter. (Present step)

SEQ LENGTH: The preset value of the counter. (Number of steps)

WORDS PER STEP: This is the width of the sequencer table.

FILE: Starting address of the sequencer table.

MASK: Address of the mask file.

Figure 8-7. Entries for an Allen Bradley PLC-2.

The mask values are used to enable and disable certain inputs and outputs. A mask is a means of selectively screening out data. Study Figure 8-8. The mask value will be used for each step. Look at step 1; it contains all ones. This means that step 1 says that all outputs should be on for both words. The mask value must be considered, however. The mask value and the step bits are "ANDed." The result becomes the actual output status for each bit. Only if the mask bit *AND* the step bit is a 1, will the actual output be turned on. One reason for this might be that each bit in the step might be a processing station in a system. The mask value might be the status of presence sensors for each station. We would only want to process the stations that had parts present. The mask value would allow us to do that.

Word 1	Word 2	
0 0 0 0 1 1 1 1 0 0 0 0 1 1 1 1	0 0 0 0 0 0 0 0 1 1 1 1 1 1 1 1	Mask Value

Word 1	Word 2	Step Number
1 1 1 1 1 1 1 1 1 1 1 1 1 1 1 1	1 1 1 1 1 1 1 1 1 1 1 1 1 1 1 1	1
1 0 0 1 0 0 0 1 1 1 1 1 1 0 0 1	1 0 0 0 0 1 1 1 1 0 0 0 0 0 0 1 1	2
0 0 1 0 1 0 0 0 0 0 1 1 1 0 0 0	0 0 0 0 0 0 0 0 0 0 0 0 1 1 1 0	3
0 1 1 0 0 0 0 0 0 0 0 0 0 1 1 1	1 1 1 1 1 0 0 0 0 0 1 0 1 0 1	ETC.

Actual Output Word 12	Actual Output Word 13	Step Number
0 0 0 0 1 1 1 1 0 0 0 0 1 1 1 1	0 0 0 0 0 0 0 0 1 1 1 1 1 1 1 1	1
0 0 0 0 0 0 0 1 0 0 0 0 0 0 1 1	0 0 0 0 0 0 0 0 0 0 0 0 0 0 1 1	2
0 0 0 0 1 0 0 0 0 0 0 0 0 1 0 0	0 0 0 0 0 0 0 0 0 0 0 0 1 1 1 0	3
0 0 0 0 0 0 0 0 0 0 0 0 0 1 1 0	0 0 0 0 0 0 0 0 0 0 0 1 0 1 0 1	Etc.

Figure 8-8. Use of a mask to determine whether or not an output is active. Note that a 1 in the mask bit means that the output is active. If the output word bit is a 1 and the mask bit is a 1, the output bit is set.

The real power of these instructions can be seen when the sequencer input and sequencer output instructions are used together. Figure 8-9 shows how a sequencer input and a sequencer output instruction might be used in a ladder diagram. The programmer can put values into the input, output, and mask tables or a sequencer load instruction can be used to place data in the sequencer file. The sequencer load instruction is programmed as an output instruction. A false-to-true transition enables this instruction. It would then place its data into a sequencer file. This could be used so that one ladder diagram program could be used to produce different products. The proper data could be loaded into the sequencer files when needed.

Sequencers are very powerful tools when programming sequential type processes. There are some limitations, however. Two events cannot run simultaneously and if one sequencer instruction is out of order, the process stops.

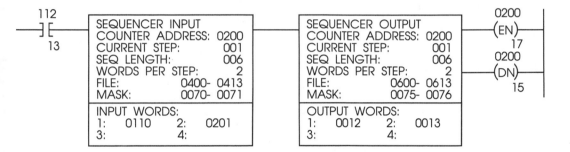

Figure 8-9. Sequencer input and sequencer output instruction used in a ladder diagram. Note that input I1213 is used to increment the sequencer input instruction. When the input sequencer instruction is true it is used to increment the output sequencer instruction. The output sequencer instruction then increments to the next step and sends that step's output conditions to the actual outputs.

Shift Resister Programming

A shift register is a storage location in memory. These storage locations can typically hold sixteen bits of data, that is, 1's or 0's. Each 1 or 0 could be used to represent good or bad parts, presence or absence of parts, or the status of outputs (see Figure 8-10). Many manufacturing processes are very linear in nature. Imagine a bottling line. The bottles are cleaned, filled, capped, and so on. This is a very linear process. There are sensors along the way to sense for the presence of a bottle, there are sensors to check fill, and so on. All of these conditions could easily be represented by 1's and 0's.

Shift registers essentially shift bits through a register to control I/O. Think of the bottling line. There are many processing stations, each represented by a bit in the shift register. We want to run the processing station only if there are parts present. As the bottles enter the line a one is entered into the first bit. Processing takes place. The stations then release their product and each moves to the next station. The shift register also increments. Each bit is shifted one position. Processing takes place again. Each time a product enters the system a 1 is placed in the first bit.

The 1 follows the part all the way through production to make sure that each station processes it as it moves through the line. Shift register programming is very applicable to linear processes.

Station #	1	2	3	4	5	6	7	8
Part Present	1	0	0	0	1	0	0	1

Figure 8-10. What a shift register might look like when monitoring whether or not parts are present at processing stations. A 1 in the station location could then be used to turn an output on that would cause the station to process material. In this case the PLC would turn on outputs at station 1, 5, and 8. After processing has occurred, all bits would be shifted to the right (in this case). A new one or a zero would be loaded into the first bit depending on whether a part was or was not present. The PLC would then turn on any stations that had a 1 in their bit. In this way processing only occurs where parts are present.

Stage Programming

Stage programming is a new concept in PLC programming. The concept is to make programming complex systems easier. This concept involves breaking the program into logical steps or stages. The stages can then be programmed individually without concern for how they will affect the rest of the program. This method of programming can reduce programming time by up to about 70 percent. It can also drastically reduce the troubleshooting time by up to 85 percent. Much of the time and effort in writing a ladder logic program is spent programming interlocks to be sure that one part of the ladder does not adversely affect another. Stages help eliminate this problem.

A process or a manufacturing procedure is simply a sequence of tasks or stages. The PLC ladder diagram that we write must make sure that the process executes in the correct order. We have to design the ladder very carefully so that rungs execute only when they should. To do this we program interlocks. The process of designing the interlocks can take the majority of programming and debugging time. A ladder logic program that deals with process control may require as much as 35 percent of the ladder dedicated to interlocking. A ladder written to control a sequential process may require that up to 60 percent be devoted to interlocking.

Let's consider a simple example. The process is shown in Figure 8-11. The process involves a conveyor and press operation. Figure 8-12 shows one ladder logic solution for the process. Remember that there are as many possible ladder logic programs as there are programmers. No two programmers would write the ladder in the same way. This makes it a little difficult when we are asked to troubleshoot or modify someone else's ladder. Even this simple process requires a fairly complex program and would require considerable time to understand. Also remember that a substantial portion of this ladder is devoted to interlocking.

What if we could program in English-like, logical blocks: for example, a block for starting the process, a block for checking for part presence, a block for locking the part, and so on. Then all we would have to do is break down any process into logical steps and the steps (or stages) would be our program. Figure 8-13 shows a solution for the press process. It is much easier to understand than a ladder program.

When we turn on the PLC it will start in stage 0 (see Figure 8-14). When the start button is pushed, the PLC changes to stage 1. Stage 1 checks for part presence. When contact X2 closes, the PLC moves into stage 2. Stage 2 locks the part by turning on coil Y1. When contact X3 closes, the PLC moves into stage 3. Stage 3 turns on coil Y2, which activates the press. When the lower limit is reached, it closes contact X4 and the PLC enters stage 4. Stage 4 raises the press by turning on coil Y3. When the press reaches the upper limit, it activates the upper limit-contact X5. The PLC then moves into stage 5. Stage 5 unlocks the part by turning coil Y4 on. When the confirm unlock contact X6 closes, the PLC moves into stage 6. Stage 6 moves the conveyor by turning on coil Y5. When X7 closes the PLC will move into stage 7.

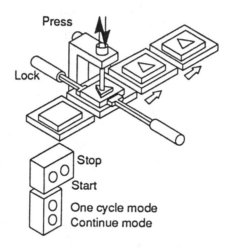

Figure 8-11. Simple press process. Courtesy Siemens Industrial Automation, Inc..

Rules of Stage Programming

Only instructions in active stages are executed. This eliminates the need for all of the complex and sometimes devious, interlocking.

Stages are activated by one of the following:

1. Power flow makes contact with a stage label.
2. A "jump to stage" is executed.
3. The stage status bit is turned on by a "set" instruction.
4. The initial stage is executed when the PLC enters run mode.

Stages are deactivated by:

a. Power flow transitions from the stage.
b. Jumping from the stage.
c. A "reset " instruction.

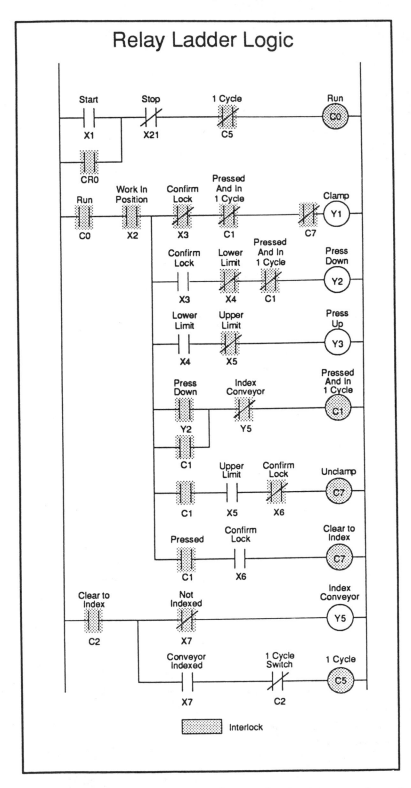

Figure 8-12. Ladder logic required to run the simple system shown in
Figure 8-11. Courtesy of Siemens Industrial Automation, Inc..

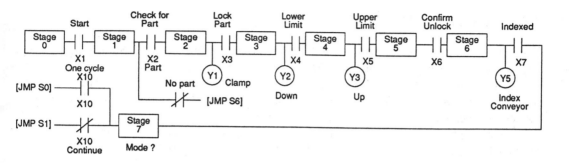

Figure 8-13. Same process as Figure 8-11 but has been programmed using stage programming. Courtesy of Siemens Industrial Automation, Inc..

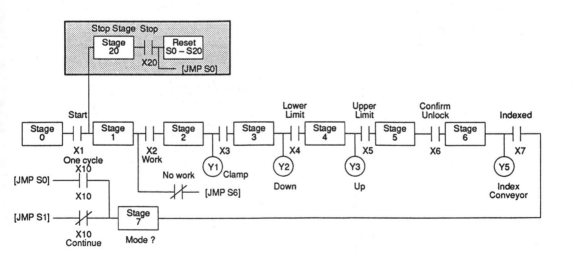

Figure 8-14. How a stop stage could be added to the stage program. Courtesy of Siemens Industrial Automation, Inc..

Let's add a stop stage to our program (see Figure 8-14). Note that now when contact X1 is closed, the PLC enters stage 1 and stage 10. Both stages are active. In fact, stage 10 will be active during all of the stages. This allows us to stop the program at any point during operation. If X20 is then closed, stages 0 through 7 are reset and the PLC is told to jump to stage 0. Stage 10 is deactivated when the PLC is in step 0.

Step Programming

Omron PLCs have a similar type of programming available. Omron calls it "Step" programming. It is very similar to stage programming. The application to be written is broken into logical steps. This is very much the way that manufacturing processes actually function. The production of a product can be broken down into logical processing steps: step A, then step B, then step C "or" step D, then step E, and so on. The beauty of step programming is that only the desired steps are active at that time. That means that there is almost no interlocking required. The individual blocks (steps) can be written and will not affect other steps.

Figure 8-15. Two types of step instruction.

There are two types of instruction blocks: step and step next (SNXT) (see Figure 8-15). The step next is used to move from one block of instructions to the next. The step instruction is used to show the beginning of a block of ladder logic. The block of instructions is then marked by a step next (SNXT) instruction.

```
0001
─┤├──────────────────────────[ SNXT HR0001 ]

              ────────────────[ STEP HR0001 ]

              ┌─────────────────────────────┐
              │     Ladder Logic for         │
              │     This Portion             │
              │     of the Process           │
              └─────────────────────────────┘
0002
─┤├──────────────────────────[ SNXT HR0002 ]
```

Figure 8-16. How a step instruction is used in a ladder diagram. Courtesy of Omron Electronics.

Input conditions are used with step next instructions to switch between blocks (steps) (see Figure 8-16). If input 0001 becomes true, the PLC will move to step HR0001. It will only evaluate the logic between step HR0001 and the following step-next instruction. It will stay in this block of instructions until input 0002 becomes true. When that happens the PLC will move to step HR0002 and evaluate the logic for that step. Note that only the logic between the step and the step next is evaluated. This eliminates all of the interlocking that normally occupies about two-thirds of the ladder and two-thirds of the programmer's time.

Figure 8-17 shows the diagram of a two-process manufacturing line. Product enters and is weighed on the input conveyor. It is then routed to one of two possible processes, depending on the weight of the product. After the individual processing all product is printed at the last processing station. Note that sensors are used to sense when product enters and leaves a process.

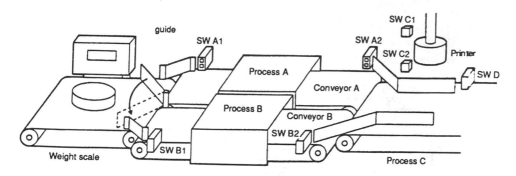

Figure 8-17. Conveyor system. The processing of the product is dependent on the weight of the product. Courtesy of Omron Electronics.

Figure 8-18 shows a block diagram of the process. Note that sensor input conditions have been used to route the product through the various possible production steps. Also note that parallel processing is possible. Steps do not have to be serial.

Figure 8-19 shows an example of what the ladder logic would look like for this process. Switches A1 and B1 have been used as input conditions to step-next instructions. If switch A1 is true, the PLC will jump to step HR0000. This is process A. The logic of step 0000 is then active until switch A2 senses the product leaving process A. Switch A2 is then true, which makes the conditions true for stepnext to jump to HR0002. The PLC then evaluates step HR0002 (process C). When the product leaves process C it makes switch D true. Switch D makes the stepnext 24614 true. This sets a bit in memory which indicates that process C is complete and has completed a part. This bit could be used by the programmer to allow further processing.

If the product weight had indicated that the product was a process B type, the logic, it would have been routed through process B and then C. While this is a relatively si? The answer would require quite a lengthy ladder diagram if it were written with no dder logic is would you rather troubleshoot, a regular ladder or a ladder written i? is that in addition to being logically organized into processing st probably one-fifth to two-thirds shorter.

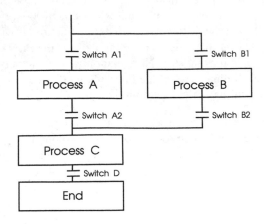

Figure 8-18. This figure shows a block diagram of the conveyor process-ing system, including the input conditions. Courtesy of Omron Electronics.

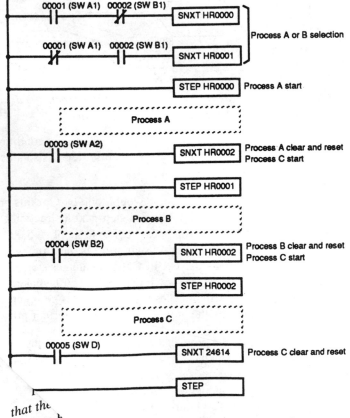

that the
Omron Ladder diagram for the conveyor processing system. Note
logic for each process has not been shown. Courtesy of

Fuzzy Logic

Fuzzy logic is a control method that attempts to make decisions like a human being would. When we make decisions we consider all of the data we have available to us based on the rules we have formulated and the present conditions. We do not use hard and fast rules, we are able to weight each rule as to its importance. This means that we do not use one fixed mathematical formula to make our decision. Fuzzy logic is an attempt to mimic human decision making. The example mentioned earlier in the book is the video recorder. Video recorders with fuzzy logic are able to differentiate wanted movement of the camera from unwanted movement and thus stabilize the picture. There are many industrial applications that are very appropriate for fuzzy logic.

The basis of fuzzy logic is the "fuzzy" set. If we thought of people's height or weight, we could easily see an image of normal height and weight. To us it seems straightforward. If you think about it though, what is normal? A bell curve can be used to show the relationship of people's heights (see Figure 8-20). Let's assume for the sake of discussion that 5 feet 9 inches is a normal height. We are not trying to say that someone who is 5 feet 1 inches or 6 feet 5 inches is not of normal height, however. Let's assume that we can readily agree that below 5 feet 0 inches is short and above 6 feet 6 inches is tall. As you can see, the definition of normal height is actually quite complex.

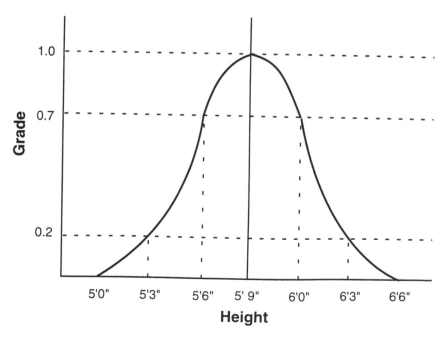

Figure 8-20. Chart of people's heights. Courtesy of Omron Electronics.

Consider Figure 8-20 again. A grade between 1.0 and 0.0 has been assigned to show how strongly we feel that the height is "normal." For example, 5 feet 9 inches has been assigned a 1.0. We do not feel quite as strongly that 5 feet 3 inches is normal height. We only assign a value of 0.2 to it. The graph (bell curve) can also be called a <u>membership</u> <u>function</u>.

Fuzzy logic decision making can be divided into two steps: the inference step and the "defuzzifier" step (see Figure 8-21).

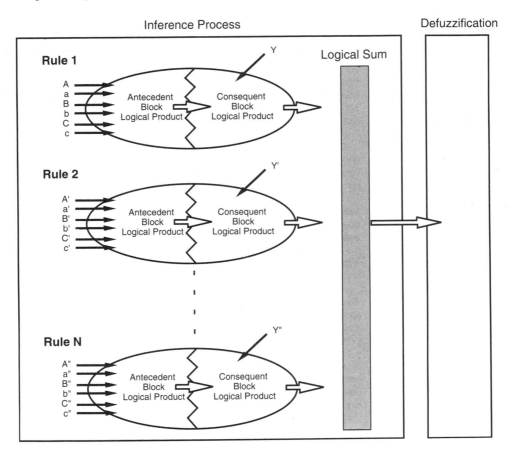

Figure 8-21. Basic fuzzy logic decision-making process. The rules are shown on the left. Decisions are made based on the individual rules. These decisions are then summed to provide a logical sum. The sum is then "defuzzed" and used to control the system. Courtesy of Omron Electronics.

The inference process is made up of several rule-based decisions that are summed to a single logical sum. Each rule is composed of input conditions (antecedent block) and a conclusion (consequent block). The logical products of each of these are summed.

Each rule is independent. They are analyzed independently. They are separate until they are combined to produce the logical sum. This combination of simple rules allows very complex decisions to be made. The rules are analyzed in parallel and a logical sum is generated. This produces a much better result than a simple formula to control all situations.

Chapter 8: Advanced Programming

Imagine a cart with a stick (see Figure 8-22). The cart must be moved to balance the stick. Earlier when we thought about height we used terms such as <u>short</u>, <u>normal</u>, and <u>tall</u>. We could call these words <u>codes</u> or <u>labels</u>. We use these labels to express degree, such as moderate, almost, or a little. This example will utilize seven labels to express degree. More or fewer labels can be used.

Figure 8-23 will at first seem complex. It is not. The figure shows the antecedent membership functions and their labels. Note that the figure actually shows a series of triangle shapes. A triangular membership function is used instead of the bell curve we used in the height example. Originally the bell-shaped membership function was used, but due to the complexity of calculation the triangular function is now used most often. The results of both are very comparable. These triangular membership functions (antecesdent membership) are composed so that the labels overlap. This permits reliable readings even when the level is not distinct or when the input from sensors is continually changing.

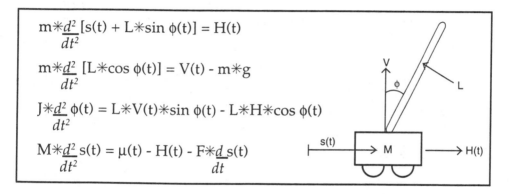

Figure 8-22. System controlled in this example. The cart must be moved just the right amount and at the right velocity to balance the stick. Diagram and example courtesy of Omron Electronics.

Developing the Rules in Code

The inclination of the stick from vertical is θ and the speed with which the inclination is changing (the angular velocity) is $\underline{d\theta}$. Both θ and $\underline{d\theta}$ are inputs from sensors. The sensors might be an encoder and a tachometer. The tachometer would give input on the angular velocity and be an encoder and a tachometer. The tachometer would give input on the angular velocity and the encoder would give input on the inclination.

These two variables can be used to write the "production" rules. The change in speed of the platform on which the stick is mounted is $\Delta \underline{V}$.

The antecedent block is usually composed of more than one variable linked by "and's." The rules are linked by "or's." In this example we use seven rules. Refer to Figure 8-23. There are seven possible states for each input (NL: negative large, NM: negative medium, etc.).

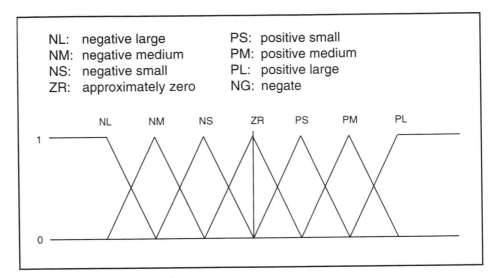

NL: negative large   PS: positive small
NM: negative medium   PM: positive medium
NS: negative small   PL: positive large
ZR: approximately zero NG: negate

Figure 8-23. This figure shows the triangular membership functions used to represent this system. Note that there are seven triangular functions shown.

Rule	Antecedent Block	Consequent Block
Rule 1	If the stick is inclined moderately to the left and is almost still	then move the hand to the left quickly
Rule 2	If the stick is inclined a little to the left and is falling slowly	then move the hand to the left slowly
Rule 3	If the stick is inclined a little to the left and is rising slowly	then keep the hand as it is
Rule 4	If the stick is inclined moderately to the right and is almost still	then move the hand to the right quickly
Rule 5	If the stick is inclined a little to the right and is falling slowly	then move the hand moderately to the right slowly
Rule 6	If the stick is inclined a little to the right and is rising slowly	then keep the hand as it is
Rule 7	If the stick is almost vertical and is almost still	then keep the hand as it is

Figure 8-24. This figure shows the 7 rules that will be used for the decision-making process. Note the consequent blocks. The consequent block shows what should be done if the rule is true.

In our example there are two inputs: inclination and angular velocity. This means that if there are seven rules, we could have 49 potential combinations. (The number of possible combinations was calculated by raising 7 to the second power.) Each combination could be a rule. For this example seven rules are used. Only important rules need to be defined.

Examine Figure 8-24. Only five of the seven states describe the angle of inclination. Only three describe the angular velocity. This reduces the possible combinations to 15 (3X5) antecedent blocks. Of the 15 remaining combinations only seven are useful in describing how the system must operate. This is very similar to the way we think. We discard the rules that are not applicable to a particular decision and use only the relevant ones.

The following explanation of fuzzy logic evaluation will at first seem very confusing. You must study the graphs carefully as you read to gain an understanding of the process. Then it will seem simple and straightforward. Study Figure 8-26. This figures show the two variables (inclination and angular velocity) at one point in time. The input value at that precise point in time is shown by the vertical line.

Each rule evaluates the input based on its membership function and assigns a value. For example, rule 1 evaluates the input based on the PM label (Positively medium inclination to the right.) The rule assigns a value of 0.7 based on where the input intersects the triangular membership function (see Figure 8-25). Rule 1 is also evaluated for the angular velocity and a value of 0.8 is assigned.

The values of rule 1 inclination and rule 1 angular velocity are then evaluated to find the logical product (minimum). This means that the minimum value is used. In this case rule 1 inclination was 0.7 and rule 1 angular velocity was 0.8. The logical product is the smallest value, or 0.7 for rule 1. Note that each of the seven rules is evaluated based on the two inputs (inclination and angular velocity). Logical minimums are found for each rule. Note that each rule is evaluated based on where the input intersects (or does not intersect) its membership function.

The next step is to find the logical sum. The logical sum is the combination of the results of the rule evaluations. Study Figure 8-26. This figure shows how the logical sum is derived. Again it may look complex at first. Study rule 1. The logical product of rule 1 was 0.7. The area equating to 0.7 is filled in the triangular membership function for the rule 1 consequent block. The result of this first evaluation would say that the cart should be moved moderately to the left quickly. Rule 2 shows that the cart should be moved to the left a little quickly (value of 0.2). The products of rules 3 to 7 were zero, so they do not affect the outcome.

The logical sum of the consequent blocks is shown on the bottom of the figure. The membership functions for rules 1 and 2 are simply combined to give the result shown at the bottom of the figure. A decision must now be made on how to move the cart (how far and how fast). This is done by calculating the center of gravity for the logical sum of the rules. The result becomes the output value.

As you have seen, seven rules and two inputs were used for this simple example. The evaluation that was done resulted in an output value that was based on an evaluation of the rules and the status of the inputs at that particular time. The advantage of fuzzy logic is that it is more flexible than mathematical models of systems. PID is a mathematical model. With the wide variety of industrial systems and operating conditions, it is very difficult to develop an accurate mathematical model for industrial systems.

Fuzzy logic helps solve the problem. Fuzzy logic breaks any system down into more humanlike rules. These simple rules are evaluated and compensate for the actual conditions at that time. Fuzzy is also easier to understand because the operator's own knowledge and thought process is reflected in the control process. PID, on the other hand, is often difficult for people to understand and adjust.

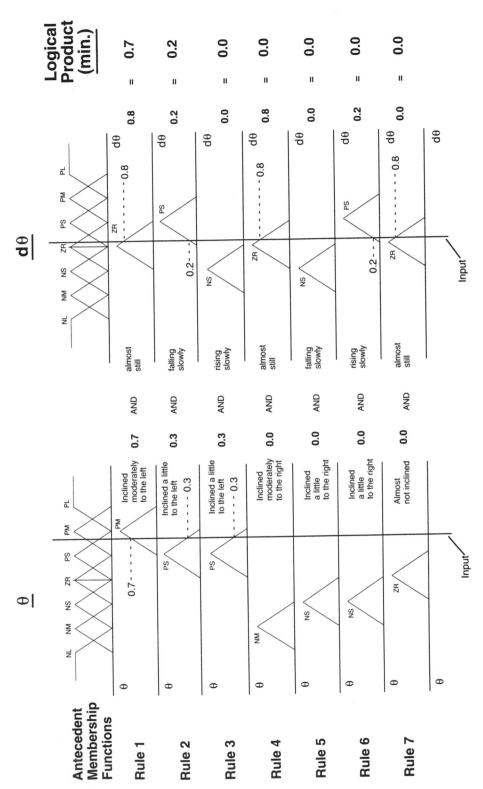

Figure 8-25. This figure shows how the two variables were evaluated by each rule at one point in time. Courtesy of Omron Electronics.

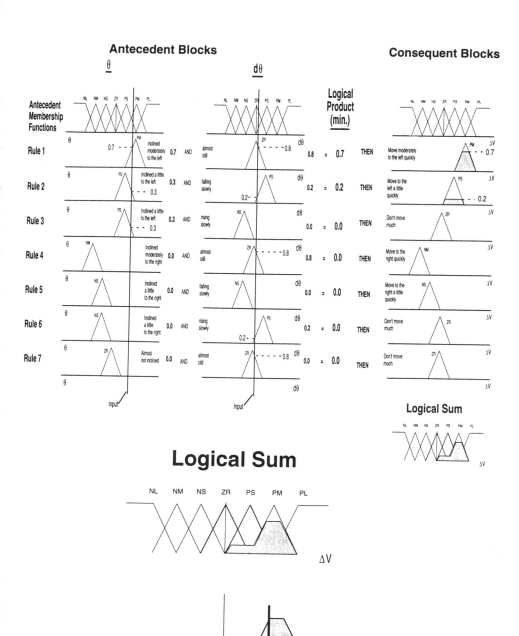

Logical Sum

Figure 8-26. *This figure shows how the logical sum is derived.*

Fuzzy theory was born about 25 years ago. In 1965, Professor L. A. Zadeh of the University of California-Berkeley, presented a paper that outlined fuzzy theory. At first the theory was met with indifference and even hostility.

In about 1970 fuzzy logic began to be used in Japan, Europe, and China. It began to show positive results. There will be room for fuzzy logic and other control systems, such as PID. Each will have it's own niche. The future of fuzzy logic does appear bright, however. Although fuzzy technology is still in its infancy, it has already contributed significantly to commercial and industrial control applications. Fuzzy technology can be applied through software, dedicated controllers, or through fuzzy microprocessors in products. This flexibility and simplicity of fuzzy logic may make fuzzy logic a basic control and information-processing technology in the twenty-first century.

Here are a few examples of applications that could benefit from fuzzy logic control:

> *Nonlinear systems such as process control, tension control, and position control.*

> *Systems with gross input deviations or insufficient input resolution.*

> *Difficult-to-control systems that require human intuition and judgment.*

> *Systems that require adaptive signal processing to overcome changing environmental or process conditions.*

> *Processes that must balance multiple inputs or that have conflicting constraints.*

State Logic

State logic is one of many new approaches to programming manufacturing systems. There are many new languages being developed to program control systems. They are an attempt to make programming an easier task. They do not use ladder logic. The commands typically used in these new languages are very straightforward English statements. State logic is presented here as an example of one of these emerging languages. State logic represents a different approach to control logic. State logic does not use ladder logic. State logic breaks processes into states and tasks. The actual logic is written in plain English. The resulting program is very easy to understand.

Study Figures 8-27 and 8-28. The system is shown in Figure 8-27 and the logic is shown in Figure 8-28. Note that the logic consists of small sections called tasks. Each task is one logical portion of the entire process. Each task is then divided into states that control a portion of the task. Each state has a statement associated with it to make decisions based on real inputs and or values and control outputs. Note how English-like the statements are.

Example System (See Figure 8-27.)
Cans are filled with 2 chemicals and mixed as they move along a conveyor belt. When a can is placed on the conveyor line, it trips the can in place limit switch, and if neither the fill nor the

mix tasks are currently active, the conveyor will start. The conveyor will run until a can arrives at either the fill or mix stations, at which time the conveyor will stop and wait for another can and any filling or mixing tasks to be completed.

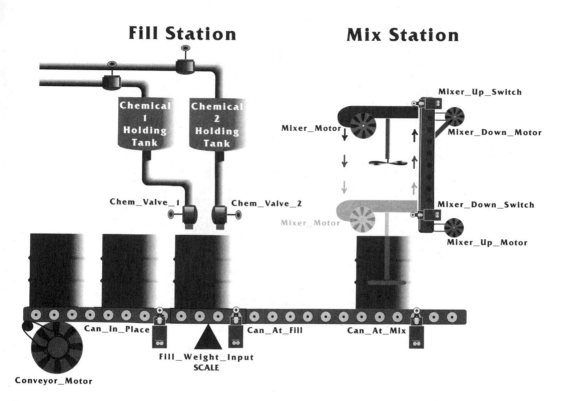

Figure 8-27 Example fill and mix system. Figure and description courtesy of Adatek.

When a can arrives at the fill station, it trips the can at fill limit switch. Chemical valve 1 will open until the fill weight is above 20 lbs. Then chemical valve 2 will open until the fill weight is above 30 lbs. Then the fill station will wait for another can to arrive to begin another cycle.

When a can arrives at the mix station, it trips the can at mix limit switch, the down mixer motor will run until the mixer down limit switch is tripped. The mixer motor will then start. After the mixer motor has run for 30 seconds the mixer up motor will run until the mixer up switch is tripped. A counter is incremented to keep track of the number of completed cans. (The counter is called Can_Inventory.) Then the message "Batch Cycle Complete" is written to the operator panel every time a can is mixed. The actual logic (program) is shown in figure 8-28.

State logic is a very high level programming language for programming systems. It is based on finite state machine theory. State logic is essentially a framework for modeling and real-world process. It is a language designed to control systems.

Project: Batching System

Task: Fill_Station
    State: PowerUp
      When Can_At_Fill is on, go to the Batch_Chem_1 State.
    State: Batch_Chem_1
      Open Chem_Valve_1
      When Fill_Weight is above 20 pounds, go to Batch_Chem_2.
    State: Batch_Chem_2
      Open Chem_Valve_2 until Fill_Weight is more than 30 lbs,
      then go to the Batch_Complete State.
    State: Batch_Complete
      When Can_At_Fill is off, go to the PowerUp State.

Task: Mix_Station
    State: PowerUp
      If Can_At_Mix is on, go to the Lower_Mixer State
    State: Lower_Mixer
      Run the Mixer_Down_Motor until the Mixer_Down_Switch is tripped,
      then go to the Mix_Chemicals State.
    State: Mix_Chemicals
      Start the Mixer_Motor.
      When 30 seconds have passed, go to the Raise_Mixer State
    State: Raise_Mixer
      Run the Mixer_Up_Motor until the Mixer_Up Switch is tripped,
      then go to the Batch_Complete.
    State: Batch_Complete
      When Can-At-Mix is off, go to the Update_Inventory State.
    State: Update_Inventory
      Add 1 to Can_Inventory.
      Write "Batch Cycle Complete" to the Operator_Panel and go to PowerUp.

Task: Conveyor
    State: PowerUp
      When Can_In_Place is on and
      (Fill_Station Task is in the PowerUp or Batch_Complete) and
      (Mix_Station Task is in the PowerUp or Batch_Complete),
      go to Start_Cycle
    State: Start_Cycle
      Start the Conveyor_Motor.
      When Can_In_Place is off, Go to Index_Conveyor State.
    State: Index_Conveyor
      Run Conveyor_Motor.
      When Can_At_Fill or Can_At_Mix is on, go to the PowerUp State.

Figure 8-28. This figure shows the actual logic for the system shown in Figure 8-27. The State Logic Control language can presently be used on personal computers and some PLCs. Logic and explanation courtesy of Adatek.

The State Logic Model

All real-world processes move through sequences of states as they operate. Every machine or process is a collection of real physical devices. The activity of any device can be described as a sequence of steps in relation to time. For example, a cylinder can exist in only one of three states: extending, retracting, or at rest. Any desired action for that cylinder can be expressed as a sequence of these three states. Even a continuous process goes through start-up, manual, run, and shutdown phases. All physical activity can be described in this manner. It is not difficult to express an event or condition that could be used to cause the cylinder (or other device) to change states. For example: If the temperature is over 100 degrees F, turn the warning light on and go to the shutdown procedure. Time and sequence are natural dimensions of the state model just as they are natural dimensions of the design and operation of every control system, process, machine, and system. State logic control uses these attributes (time and sequence) as the components of program development. As a result, the control program is a clear snapshot of the system that is being controlled. State logic is a hierarchical programming system that consists of tasks, states, and statements.

Tasks

Tasks are the primary structural element of a state logic program. A task is a description of a process activity expressed sequentially and in relation to time. If we were describing an automobile engine the tasks would include the starting system task, the fuel system task, the charging system task, the electrical system task, and so on. Almost all processes contain multiple tasks operating in parallel. Tasks operating in parallel is necessary because most machines and processes must do more than one thing at a time. State logic provides for the programming of many tasks that are mutually exclusive in activity yet interactive and joined in time.

States

States are the building blocks of tasks. The activity of a task is described as a series of steps called states. A state describes the status or value of an output or group of outputs. These are the outputs of the control system and thus are inputs to the process. Every state contains the rules which allow the task to transition to another state. A state is a subset of a task that describes the output status and the conditions under which the task or process will change to another state. The states when taken in aggregate provide a description of the sequence of activity of the process or machine under control. Further it provides an unambiguous specification of how that portion of the process will respond in all conditions.

Statements

Statements are the user's command set to create state descriptions. The desired output related activity of each state can be described by using statements. Statements can initiate actions or can base an output status change on a conditional statement or a combination of conditions. Any input value or variable can be used in conditional statements. Variables can include state status from other tasks as well as typical integer, time, string, analog, and digital status variables.

State logic allows the programmer to write the control program in natural English statements. State logic and other types of system languages are sure to increase in acceptance and popularity because of their simplicity and ease of use.

Questions

1. What is a drum controller?

2. What are the benefits of a drum controller?

3. List at least three disadvantages of drum controllers.

4. How is a sequencing instruction different from a drum controller?

5. Construct a matrix of what a stoplight program would look like. Show the east/west outputs and north/south outputs as well as the steps.

6. Explain how the matrix you constructed in question 5 would be used to program a sequencer.

7. What is stage/step programming?

8. List at least three advantages of stage/step programming.

9. What is a shift register?

10. What types of applications are appropriate for the use of shift register programming?

11. What is fuzzy logic?

12. What are rules?

13. What types of applications are appropriate for fuzzy logic?

14. What is state logic programming?

Additional Exercises

1. Study the Siemens Industrial Automation, Inc. stage tutorial software. It is available from your instructor.

    a. What are the rules for transitioning from one stage to another?

    b. Can two stages be active at the same time?

    c. How can a stage be deactivated?

2. Study the ECLIPS (English Control Language Programming System) state logic tutorial. It is available from your instructor. The tutorial will show you how a system can be programmed and debugged using state logic.

Chapter 9

Overview of Plant Floor Communication

If enterprises are to become more productive, they need to improve processes. This requires accurate data. Communications are vital. Production devices hold very valuable data about their processes. In this chapter we examine how these data can be acquired.

Objectives

Upon completion of this chapter, the student will be able to:

Describe the four levels of plant communications and characteristics at each level.

Compare and contrast human communications and machine communications.

Define such terms as serial, synchronous, RS-232, RS-422, device, cell, area, host, and SCADA.

Describe how computers can communicate with PLCs.

Explain the concept of MAP/TOP.

Describe the opportunity available through factory communications.

Introduction

The programmable controller has revolutionized manufacturing. It has made automation flexible and affordable. PLCs control processes across the plant floor. In addition to producing product, PLCs also produce data. The data can be more profitable than the product. This may not seem obvious; however, most processes are very inefficient. If we can use the data to improve processes we can drastically improve profitability. The inefficiencies are not normally addressed because people are busy. There are other more pressing problems. (A good friend of mine in a small manufacturing facility said it best. "It's hard to think about fire prevention when you're in the middle of a forest fire.") Manufacturing people are usually amazed to find that most of the data they would like to have about processes is already being produced in the PLC. With very few changes the data can be gathered and used to improve quality, productivity, and uptime. There are huge gains possible if these data are used.

To be used the data must first be acquired. Many managers today will say that they are already collecting much of the data. Why should they invest in electronic communications when they are already gathering data from the plant floor manually? The reasons are many. Often, data gathered manually are very inaccurate. The data are not real-time either. The information must be written down by an operator, gathered by a foreman, taken "upstairs", entered by a data processing person, printed into a report, and distributed.

This all takes time. It can often make data many days or weeks late. If mistakes in entry were made on the floor, it is often too late to correct. The reports that are produced are often mazes of meaningless information: too much extraneous information to be useful to anyone. The lateness and inaccuracy of the data gathering makes it almost counterproductive.

The other communication that is required is realtime information to the operator: accurate orders, accurate instructions, current specifications, and so on. This is often lacking in industrial and service enterprises today.

This communication is quite easily achieved with the use of electronic communications. Many of the data required already exist in the smart devices on the factory floor. Much of the data that people write on forms in daily production already exist in the PLC.

The improvements in computer hardware and software have made communication much easier. There are many software communications packages that make it easy to communicate with PLCs. Communications will increase in importance as American manufacturing tries to improve its competitive position.

Levels of Plant Communication

Plant floor communications can be broken down into levels. Some authors break them into four levels, some into five. The basic concepts are the same. In this book we use a four-level model. The four levels are: device, cell, area, and host (see Figure 9-1). Each level is a vital link. The device level is the production level. As we move up the pyramid, management of production becomes the task.

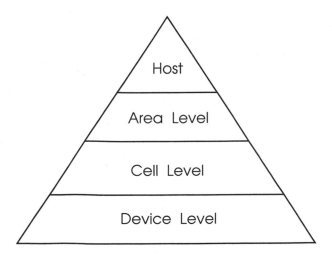

Figure 9-1. Four-level model of plant communication.

The easiest way to understand this model is to relate it to people. Figure 9-2 shows the human model of factory communications. The workers are at the device level. The foreman is at the cell level. The factory supervisors are at the area level, and the plant manager is at the host level. If we think of the typical duties that each person would have at each level it is easy to understand the function of each level in the electronic model. Each of these will be covered in more detail.

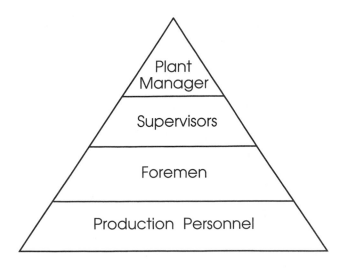

Figure 9-2. Comparison of the plant communication model and the typical industrial organization of people. It is necessary to communicate information down to the workers and also to send information back up to the top of the organization.

Device Level

The device level is the lowest level of control on the plant floor. Think of devices as pieces of equipment that produce or handle product (see Figure 9-3). Some examples of devices are robots, conveyors, computer-controlled machine tools (CNC machines), hard automation, and automated storage/automated retrieval systems (AS/AR). These devices have several things in common. They are all directly involved with the product. Some move the product, others add value during production. Each device produces valuable data during production. They know how many have been produced, how long it takes to produce the product (cycle time), uptime and downtime, and so on.

In the human model the device level relates to production personnel. Production personnel add value to the product. They handle the product and add value in some manner. They also fill out paperwork so that management can monitor quality, productivity, and so on.

In the electronic model the device might be a CNC machine. The CNC produces machined parts very efficiently. In addition to producing parts, the CNC is also producing data. The CNC can track cycle times, piece counts, downtime, etc. If it were used, the data could drastically improve productivity.

Unfortunately, very few manufacturers use the data. In the rest of the cases the data invisibly and continually spill out the back of the device onto the floor or into the infamous bit bucket. The bit bucket is a fictitious object. The fact is that data generated in devices are not used. This is a huge lost opportunity.

One of the other things that devices have in common is a need for data. Each device needs a program to tell it what, when, and how many to produce. The programs for each will look different for each device, but serve the same function. In the human model, people need the same information. What should I work on? How many should I make? What should I do next? Is there enough raw material? Etc., Etc., Etc. This information is crucial to efficient production. Unfortunately, inaccurate, untimely information is more the rule than the exception in enterprises. In most factories, people are used to coordinate the devices. People start and stop the machines, count pieces, monitor quality, monitor performance, and watch for problems. The importance of accurate, timely information is just as vital when people are involved.

Figure 9-3. Device level. Devices are production-oriented equipment that add value to a product.

When you think of the device level think of task-specific equipment that is adding value to the product. Think of production-type tasks. The device level is where value is added and thus where the enterprise makes its money. It is this level that creates the wealth to support the rest of the organization. Some may be uncomfortable with that thought. The Japanese view the production level as the most important. They do anything they can to make that level more efficient. They use the concept of <u>Kaizen,</u> or continuous improvement. Accurate, real-time data can be crucial in real improvement.

Chapter 9: Overview of Plant Floor Communication

Cell Level

The cell is a logical grouping of devices used to add value to one or more product. A cell will typically work on a family of similar parts. A cell consists of various dissimilar devices (see Figure 9-4). Each device typically has its own unique type of program and communication protocol. Devices do not want to communicate with each other. A CNC machine, for example, does not want to communicate with a robot.

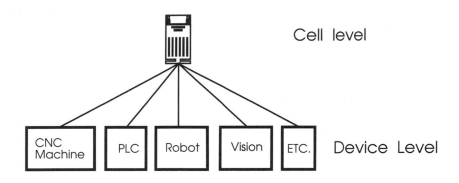

Figure 9-4. Typical cell control scheme. One cell controller is controlling several devices.

Compare this to the human model. The foreman is the cell controller. The foreman's job is to show up early for work and find out what needs to be produced that day. The foreman must then choose the appropriate production person to perform each task that needs to be accomplished. Note that all people have different personalities. A good foreman can communicate well with each individual. The foreman then monitors each employee's performance and by coordinating efforts produces that days product. The foreman makes sure that people cooperate to get the overall job accomplished.

The purpose of the cell controller is to integrate the various devices into a cooperative work cell. The cell controller must then be able to communicate with each device in the cell. Even if the devices are not able to communicate with each other, they must be able to communicate with the cell controller. The cell controller must be able to upload/download programs, exchange variable information, start/stop the device, and monitor the performance of each device.

There can be many foremen in a plant, and there can be many cell controllers in a plant. Each can control a group of devices. Each cell controller can talk to other cell controllers and also up to the next level. There are two main types of communication: primitive and complex.

Primitive Communications

Some devices do not have the capability to communicate. Some simple PLCs , for example, cannot communicate serially with other devices. In this case primitive methods are used. In the primitive mode the devices essentially just handshake with a few digital inputs and outputs. For example, a robot is programmed to wait until input 7 comes true before executing program number 13. It is also programmed to turn on output 1 after it completes the program. A PLC output can then be connected to input 1 of the robot and output 1 from the robot can be connected to an input of the PLC. Now we have a simple one-device cell with primitive communications. The PLC can command the robot to execute. When the program is complete, the robot will notify the PLC. Note that it is very simple yes or no (binary) information.

Many devices offer more communications capability. For example, we may need to upload/download programs or update variables. We cannot do that with primitive communications. Most devices offer serial communications capability, using the asynchronous communications mode and have a RS-232 serial port available. One would think that any device with a RS-232 port would easily communicate with any other device with a RS-232 port. This is definitely not true. Devices have their own protocol.

The RS-232 standard specifies a function for each of 25 pins. It does not say that any of the pins must be used however. Some manufacturers use only three, as in Figure 9-5. Some device manufacturers use more than three pins, so some electrical handshaking can take place. In Figure 9-5 no handshaking is taking place. The first computer sends a message whether or not there is another device there. The computer cable could be unplugged or the computer turned off, the sender would not know. Handshaking implies a cooperative operation. The first computer tells the second that it has a message it would like to transmit. It does this by setting pin 4 (the request to send pin) high. The second device sees the request to send pin high, and if it is ready to receive, it sets the clear to send pin high. The first computer then knows that the cable is connected, the computer is on, and it is ready to receive. Some devices can be set up to handshake; others cannot.

Fortunately when a device is purchased it is generally capable of communicating with an IBM personal computer. The user usually just has to open the device manual to the section on communication to find a pinout for the proper cable. It is still difficult and expensive to communicate when a wide variety of devices are involved.

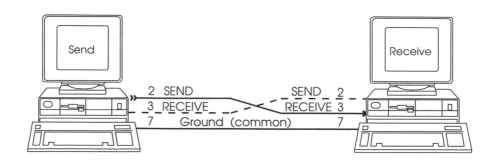

Figure 9-5. Simple RS-232 wiring scheme. It shows the simplest of RS-232 connections.

When a message is sent using asynchronous communications the message is broken into individual characters and transmitted one bit at a time. The ASCII system is used normally. In ASCII every letter, number, and some special characters have a binary-coded equivalent. There

Chapter 9: Overview of Plant Floor Communication

is 7-bit ASCII and 8-bit extended ASCII. In 7-bit ASCII there are 128 possible different letters, numbers, and special characters. In 8-bit ASCII, 256 are possible.

Each character is sent as its ASCII equivalent. For example, the letter "A" would be 1000001 in 7-bit ASCII (see Figure 9-6). It takes more 7 bits to send a character in the asynchronous model, however. There are other bits that are used to make sure the receiving device knows a message is coming, that the message was not corrupted during transmission, and even bits to let the receiver know that the character has been sent. The first bit sent is the start bit (see Figure 9-7). This lets the receiver know that a message is coming. The next 7 bits (8 if 8-bit is used) are the ASCII equivalent of the character. Then there is a bit reserved for parity. Parity is used for error checking. The parity of most devices can be set up for odd or even, mark or space, or none.

Figure 9-6. How the letter "A" would be transmitted in the asynchronous serial mode of communications. This example assumes odd parity. There are an even number of ones in the character "A," so the parity bit is a one to make the total odd. If the character to be sent has an odd number of ones, the parity bit would be a zero. The receiving device counts the number of ones in the character and checks the parity bit. If they agree, the receiver assumes that the message was received. accurately. This is rather crude error checking. Note that two or more bits could change state and the parity bit could still be correct but the message wrong.

Start bit	Data bits	Parity bit	Stop bit
1 Bit	7 or 8	1 Bit (Odd,Even, Mark, Space or None)	1, 1.5, or 2 bits

Figure 9-7. How a typical ASCII character is transmitted.

There are some new standards that will help integrate devices more easily. RS-422 and RS-423 were developed to overcome some of the weaknesses of RS-232. The distance and speed of communications are drastically higher in RS-422 and RS-423. These two standards were developed in 1977.

RS-422 is called balanced serial. RS-232 has only one common. The transmit and receive line use the same common. This can lead to noise problems. RS-422 solves this problem by having separate commons for the transmit and receive lines. This makes each line very noise immune. The balanced mode of communications exhibits lower crosstalk between signals and is less susceptible to external interference. Crosstalk is the bleeding of one signal over onto another. This reduces the potential speed and distance of communications. This is one reason that the distance and speed for RS-422 is much higher. RS-422 can be used at speeds of 10 megabits for distances of over 4000 feet, compared to 9600 baud and 50 feet for RS-232.

RS-423 is similar to RS-422 except that it is unbalanced. RS-423 has only one common which the transmit and receive lines must share. RS-423 allows cable lengths exceeding 4000 feet. It is capable of speeds up to about 100,000.

RS-449 is the standard that was developed to specify the mechanical and electrical characteristics of the RS-422 and RS-423 specifications. The standard addressed some of the weaknesses of the RS-232 specification. The RS-449 specification specifies a 37-pin connector for the main and a 9-pin connector for the secondary. Remember that the RS-232 specification does not specify what type of connectors or how many pins must be used.

These standards are intended to replace RS-232 eventually. There are so many RS-232 devices that it will take a long time. It is already occurring rapidly in industrial devices. Many PLC's standard communications are done with RS-422.

Adapters are cheap and readily available to convert RS-232 to RS-422 or vice versa (see Figure 9-8). This can be used to advantage if a long cable length is needed for a RS-232 device (see Figure 9-9).

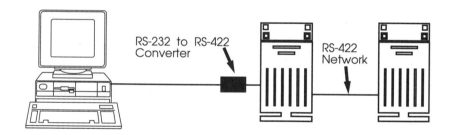

Figure 9-8. Use of a converter to change RS-232 communications to RS-422 communications. Note that the computer is then able to communicate with other devices on its network. The devices on the right are on an RS-422 network.

RS-485 is a derivation of the RS-422 standard. It is unbalanced, however. The main difference is that it is a multidrop protocol. This means that many devices can be on the same line. This requires that the devices have some intelligence, however, because the devices must each have a name so that each knows when it is being talked to.

Types of Cell Controllers

PLCs and computers can both be used for cell control applications. There are definite advantages and disadvantages for each.

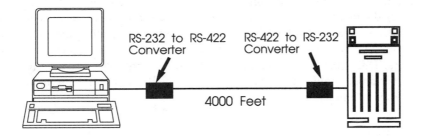

RS-232 to RS-422
Converter

RS-422 to RS-232
Converter

4000 Feet

Figure 9-9. Use of two converters. In this case it is being done to extend the cable length. Remember that RS-232 is only reliable to about 50 feet. The use of two converters allows 4000 feet to be covered by the RS-422 and then converted to RS-232 on each end. Note that the speed will be limited by the RS-232.

PLCs as Cell Controllers

The PLC offers some unique advantages as a cell controller. It is easily understood by plant electricians and technicians. If the devices in the cell need to communicate in primitive mode, it is very easy to do with a PLC. If there are other PLCs of the same brand in the cell, it is easy for the PLC to communicate. The data highway of that brand PLC would then be used.

The PLC is not very applicable as a cell controller when there is more than one brand of device in the cell. Typical PLCs do not offer as much flexibility in operator information as computers do, although this is changing rapidly. Graphic terminals and displays are becoming very common for PLCs.

Computers as Cell Controllers

The computer offers more flexibility and capability than the PLC. The computer has much more communications capability than that of any PLC. In fact, all device manufacturers want their devices to communicate with a microcomputer. It is difficult to sell a device that does not communicate with an IBM microcomputer. Most device manufacturers do not care to communicate with other brands of PLCs, robots, and so on. They all want to talk to an IBM compatible, however. This makes it much easier for the computer to act as cell controller. It can communicate with each device in the cell.

This communication is usually accomplished by the use of SCADA (supervisory control and data acquisition) software. The concept is that software is run in a common microcomputer to enable communications to a wide variety of devices. The software is typically like a generic building block (see Figure 9-10).

The programmer writes the control application from menus or in some cases graphic icons. The programmer then loads drivers for the specific devices in the application. Drivers are software. A driver is a specific package that was written to handle the communications with a specific brand and type of device. They are available for most common devices and are relatively inexpensive.

The main task of the software is to communicate easily with a wide range of brands of devices. Most enterprises do not have the expertise required to write software drivers to communicate with devices. SCADA packages simplify the task. In addition to handling the communications, SCADA software makes it possible for applications people to write the control programs instead of programmers. This allows the people who best know the application to write it without learning complex programming languages. Drivers are available for all major brands of PLCs and other common manufacturing devices.

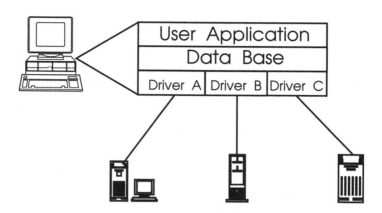

Figure 9-10. How a typical SCADA software package works. Note that the user application defines which variables from the devices must be communicated. These are collected through the drivers and stored in a database that is available to the application. Once the computer has the desired data, it is a relatively easy task to make the data available to other devices.

In general, an applications person would write the specific application using menu-driven software. The software is easy to use. Some are like spreadsheets and some use icons for programming. Instead of specific I/O numbers that the PLC uses, the programmer uses tag-names.

For example, the application might involve temperature control. The actual temperature might be stored in register S20 in the PLC. The applications programmer would use a tagname instead of the actual number. The tagname might be "temp1."

This makes the programming transparent. Transparent means that the application programmer does not have to worry very much about what brand of devices are in the application. A table is set up that assigns specific PLC addresses to the tagnames (see Figure 9-11). In theory, if a different brand PLC were installed in the application, the only change required would be a change to the tagname table and the driver.

Fortunately, there is more software available daily to make the task of communications easier. The software is more friendly, faster, more flexible, and more graphics-oriented. The data gathered by SCADA packages can be used for statistical analysis, historical data collection, adjustment of the process, or graphical interface for the operator.

Device	Actual Number	Tagname
PLC 12	REG20	Temp1
PLC 12	REG12	Cycletime1
PLC 10	S19	Temp2
PLC 07	N7:0	Quantity1
Robot 1	R100	Quantity2

Figure 9-11. What a tagname table might look like.

Area Control

Area controllers are the supervisors (see Figure 9-12). They look at the larger picture. They receive orders from the host and then assign work to cells to accomplish the tasks. They also communicate with other area controllers to synchronize production. Area controllers use synchronous communications methods. Area controllers are attached via local area networks (LANs).

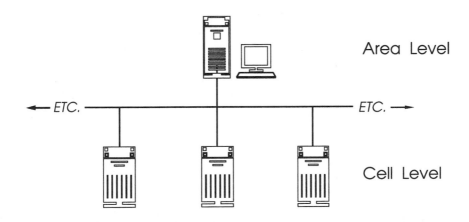

Figure 9-12. Typical area control diagram.

Local Area Networks (LANs)

Local area networks are the backbone of communications networks. The topic of LANs can be broken down into various methods of classification. We examine three: topology, cable type, and access method.

Topology

Topology refers to the physical layout of LANs. There are three main types of topology: star, bus and ring.

Star Topology

The star style uses a hub to control all communications. All nodes are connected directly to the hub node (see Figure 9-13). All transmissions must be sent to the hub, which then sends them on to the correct node. One problem with the star topology is that if the hub goes down, the entire LAN is down.

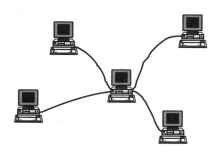

Figure 9-13. Star topology.

Bus Topology

The bus topology is a length of wire from which nodes can be tapped into. At one end of the wire is the head end (see Figure 9-14). The head end is an electronic box that performs several functions. The head end receives all communications. The head end then remodulates the signal and sends it out to all nodes on another frequency. Remodulate means that the headend changes the received signal frequency to another and sends it out for all nodes to hear. Only the nodes that are addressed pay attention to the message. The other end of the wire (bus) dissipates the signal.

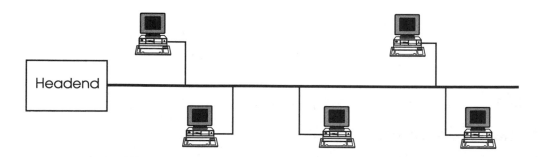

Figure 9-14. Typical bus topology. Each node (communication device) can speak on the bus. The message travels to the headend and is converted to a different frequency. It is then sent back out and every device receives the message. Only the device that the message was intended for pays attention to the message.

Ring Topology

The ring topology looks like it sounds. It has the appearance of a circle (see Figure 9-15). The output line (transmit line) from one computer goes to the input line (receive line) of the next computer, and so on. It is a very straightforward topology. If a node wants to send a message, it just sends it out on the transmit line. The message travels to the next node. If the message is addressed to it, the node writes it down; if not, it passes it on until the correct node receives it.

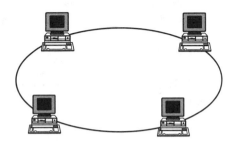

Figure 9-15. Ring topology.

The more likely configuration of a ring is as shown in Figure 9-16. This style is still a ring; it just does not appear to be. This is the convenient way to wire a ring topology. The main ring (backbone) is run around the facility and interface boxes are placed in line at convenient places around the building. These boxes are often placed in "wiring closets" close to where a group of computers will be attached. These interface boxes are called multiple station access units (MSAUs). They are just like electrical outlets. If we need to attach a computer, we just plug it into an outlet on the MSAU. The big advantage of the MSAU is that devices can be attached/ detached without disrupting the ring. Communications are not disrupted at all.

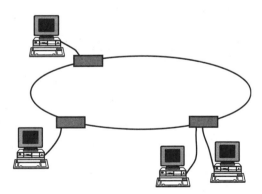

Figure 9-16. Typical ring topology. Note that it does not look like a ring. It looks more like a star topology. It actually is a ring, however. The multiple station access units (MSAUs) allow multiple nodes to be connected to the ring. The MSAU looks like a set of electrical outlets. The computers are just plugged into the MSAU and are then attached to the ring. The MSAU has relays for each port. This allows devices to be plugged in and removed without disrupting the ring.

Cable Types

There are four main types of transmission media: twisted pair, coaxial, fiber optic, and radio frequency. Each has distinct advantages and disadvantages. The capabilities of the cable types are expanding continuously.

Twisted Pair

Twisted-pair wiring is, as its name implies, pairs of conductors (wires) twisted around each other along their entire length. The twisting of the wires helps make them more noise immune. The telephone wires that enterprises have throughout their buildings are twisted pair. There are two types of twisted-pair wiring: shielded and unshielded. The shielded type has a shield around the outside of the twisted pair. This helps further to make the wiring noise immune. The wiring that is used for telephone wiring is typically unshielded. The newer types of unshielded cable are more noise immune than in the past. Higher speeds are being accomplished continuously.

Unshielded cable is very cheap and easy to install. The hope is that companies may be able to use spare telephone twisted pairs to run the LAN wiring instead of running new cable. It must be remembered though that much of the twisted-pair wire in buildings today is the older, less noise immune type.

The shielded twisted pair is now used commonly for speeds up to 16 megabits. There are companies that supply LAN cards for unshielded twisted pair that can also run at 16 megabits. There are committees studying the feasibility of up to 100 megabits for shielded twisted-pair wiring. There should soon be a standard for 100 megabits on shielded twisted-pair wiring.

Coaxial Cable

Coax is a very common communication medium. Cable TV uses coaxial cable. Coax is broadband, which means that many channels can be transmitted simultaneously. Coax has excellent noise immunity because it is shielded (see Figure 9-17).

Figure 9-17. Coaxial cable. Note the shielding around the conductor.

<u>Broadband</u> technology is more complex than <u>baseband</u> (single channel). With broadband technology there are two ends to the wire. One of the ends is called the head end. The head end receives all signals from devices that use the line. The head end then remodulates (changes to a different frequency) the signal and sends it back out on the line. All devices hear the transmission but only pay attention if it is intended for their address.

<u>Frequency-division multiplexing</u> is used in broadband technology. The transmission medium is divided into channels. Each channel has it's own unique frequency. There are also buffer frequencies between each channel to help with noise immunity. Some channels are for transmission and some are for reception.

<u>Time-division multiplexing</u> is used in the baseband transmission method. This method is also called <u>time slicing</u>. There are several devices that may wish to talk on the line (see Figure 9-18). We cannot wait for one device to finish its transmission completely before another begins. They

must share the line. One device takes a slice of time, then the next does, and so on. There are several methods of dividing the time on the line.

Figure 9-18. How time-division multiplexing works. Each device must share time on the line. Device 1 sends part of its message and then gives up the line so that another device can send, and so on.

Fiber-Optic Cable

Fiber-optic technology is also changing very rapidly. (See Figure 9-19 for the appearance of the cable.) The major arguments against fiber are: it's complexity of installation and high cost, however, the installation has become much easier, and the cost has fallen dramatically, to the point that when total cost is considered, fiber is not much different for some installations than shielded twisted pair.

Figure 9-19. Fiber optic cable. Note the multiple fibers through one cable.

The advantages of fiber are its perfect noise immunity, high security, low attenuation, and high data transmission rates possible. Fiber transmits with light, so that it is unaffected by electrical noise. The security is good because fiber does not create electrical fields that can be tapped like twisted pair or coaxial cable. The fiber must be physically cut to steal the signal. This makes it a much more secure system.

All transmission media attenuate signals. That means that the signal gets progressively weaker the farther it travels. Fiber exhibits far less attenuation than other media. Fiber can also handle far higher data transmission speeds than can other media. The FDDI (fiber distributed data interchange) standard was developed for fiber cable. It calls for speeds of 100 megabits. This seemed very fast for a short period of time, but it is thought by many that it may be possible to get 100 megabits with twisted pair, so there is now thought to raising the speed standard for fiber.

Plastic fiber cable is also gaining ground. It is cheap and easy to install. The speed of the plastic is much less but is constantly being improved.

Radio Frequency

Radio-frequency (RF) transmission has recently become very popular. The use of RF has exploded in the factory environment. The major makers of PLCs have RF modules available for their products. These modules use radio waves to transmit the data. The systems are very noise immune and perform well in industrial environments. RF is especially attractive because no wiring needs to be run. There are even RF LANs available for office environments. This is a major advantage for office areas where changes are made frequently.

Access Methods

Token Passing

In the token-passing method, only one device can talk at a time. The device must have the token to be able to use the line. The token circulates among the devices until one of them wants to use the line (see Figure 9-20). The device then grabs the token and uses the line.

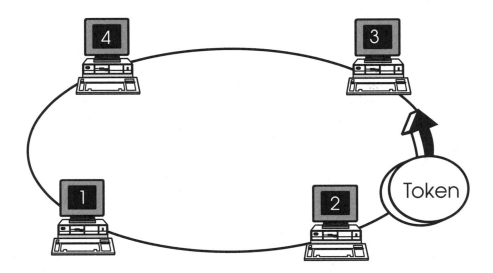

Figure 9-20. Token-passing method.

The device that would like to talk waits for a free token.

The sending station sets the token busy bit, adds an information field, adds the message it would like to send, and adds a trailer packet. The header packet contains the address of the station for which the message was intended. The entire message is then sent out on the line.

Every station examines the header and checks the address to see if it is being talked to. If not, it ignores the message.

The message arrives at the intended station and is copied. The receiving station sets bits in the trailer field to indicate that the message was received. It then regenerates the message and sends it back out on the line. The original station receives the message back and sees that the message was received. It then frees the token and sends it out for other stations to use.

Token passing offers very reliable performance. It also offers predictable access times, which can be very important in manufacturing.

Collision Detection

Collision detection is a method by which any devices that wish to speak must listen to the carrier signal on the transmission line. If the line is not busy, the device may use the line to communicate. If two or more devices try to use the line at the same time, there is a collision. The collision is detected and the devices back off for awhile and try again.

This principle is called CSMA/CD (carrier sense multiple access/collision detection). It is somewhat analogous to access to a highway. The more cars there are on the highway, the more difficult it is to get on the highway. There are also more collisions as the traffic increases. With data transmission it is far less serious. The device just retransmits after a short wait. Typical industrial communications involve low line-utilization percentages. CSMA/CD performs well at these levels.

The future of data transmission will involve much more transmission of graphics. Graphics involve very large files, which will increase line utilization levels substantially.

Manufacturing Automation Protocol

Imagine what it would be like if there were no standards for consumer goods and appliances. If you decided to switch your brand of refrigerator, you might have to add a transformer to supply the correct voltage, change outlets so that you could plug it in, and so on. Think what your house wiring would be like if every device manufacturer chose their own standard for voltage levels and connectors. It would be a nightmare! That may give you a small idea of what the situation is like in factory communications between devices. MAP (manufacturing automation protocol) was intended to solve many of the problems of connecting devices.

MAP was the idea of General Motors. GM had tens of thousands of devices in the early 1980s, very few of which were able to communicate outside their environment. Every brand and model of device had its own protocol. It would be almost impossible economically to write software drivers for each of the devices, so that all could communicate.

The idea of a standard for communications was born. The thought was that if a standard communications protocol existed and customers required it, manufacturers would produce devices that met the standard. The MAP standard is based on a standard called the open systems interconnect (OSI) model. OSI stands for open systems interconnect. It is a seven-layer method of standardizing communications among various devices (see Figures 9-21 and 9-22).

MAP had some early problems. The first standard, MAP 2.0, did not specify every layer. There also were not many devices available that were MAP compatible. The early devices were expensive. Most plants already had substantial investments in other protocols also.

In 1987, Map 3.0 was released. MAP 3.0 addressed many of the complaints and weaknesses of MAP 2.0. MAP 3.0 addresses all seven layers of the OSI model. (See Figures 9-21 and 9-22.) It only allows compatible extensions for six years. This allows manufacturers to plan ahead and not worry about an updated standard that would make their equipment obsolete. In 1992 there are approximately 20 MAP installations in the United States. Some estimates show that Japan has a similar number of installations.

The layers of the OSI model perform the following functions.

> ***Physical layer: encodes and physically transfers the message to another network device.***

> ***Link layer: maintains links between devices and performs some error checking.***

> ***Network Layer: establishes connections between devices attached to the network.***

> ***Transport layer: provides for transparent, reliable data transfer between devices.***

> ***Session layer: translates names and addresses, provides access security, and also synchronizes and manages the data.***

> ***Presentation layer: restores data to and from the standardized format that is used in the network.***

> ***Application layer: provides services to the users application program in a format that it can understand.***

OSI Seven-Layer Stack

Application
Presentation
Session
Transport
Network
Data Link
Physical

Figure 9-21. OSI model of communications.

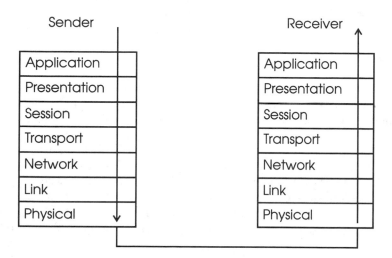

Figure 9-22. Message being sent from one device to a different device with a different protocol. The sender sends the message down through the seven layers. As it moves through the layers it is made generic, broken into transmission blocks, and put on the network and travels to the receiving device. The message moves through the stack at the receiver and the message is changed from generic to the protocol that the receiving device understands. The layers also provide error checking, security, choose the best path for the message, and so on.

Technical Office Protocol

The communication needs of the business portion of the business were seen to be different from those of the factory floor. Communication at this level is usually larger files, such as business data, engineering data, and so on. The TOP (technical office protocol) model was based on the contention access scheme. Boeing Corporation was instrumental in developing the standard.

Manufacturing Message Specification

MMS (manufacturing message specification) is a concept whose time may have come. The concept is that every device does not have to be exactly the same, to communicate. Imagine a convention of people from various countries with no common language and no interpreters. The convention would be less than successful. But what if each representative knew a core set of words that had the same meaning for each? They could get by quite well with a reduced vocabulary.

The same is true of devices. There are not too many commands we need to send to a device. We need to upload/download, start/stop, monitor, update variables, and a few more tasks. If every device could handle the commands for these few tasks in the same way they could function very well in an integrated environment. That is the principle behind MMS. There is a set of functions that has been established as a standard. As of this time there are over 80 standard functions. If a device is MMS compatible, it can use all or a subset of these functions. This will make communications a much easier task. MMS will reside above the seven-layer stack.

Host Level

The host-level controller is generally a mainframe. This computer is responsible for the business software, engineering software, office communications software, and so on. The line between mainframe, minicomputer, and personal computer is rapidly blurring. The trend is definitely toward distributed processing.

The business software is generally a MRP package. MRP stands for manufacturing resource planning. This software is used to enter orders, bills of materials, check customer credit, inventory, and so on. The software can then be used to generate work orders for manufacturing, orders for raw materials and component parts, schedules, and even customer billing. They are being used more and more for planning and forecasting. (They are typically called MRPII now because of the increased emphasis on planning and forecasting.)

The host level is going to be used more and more to optimize operation of the enterprise. Data from the factory floor will be gathered automatically from devices or from operator interface terminals. The host level's task will be to analyze the data to help improve the productivity of the overall system. The host level must also schedule and monitor the daily operations.

The real key to the future will be this data collection. Better business decisions can be made if accurate real-time, data are available in a format that people can understand and use.

Figure 9-23 shows the levels of enterprise communications. The task of integrating these devices is rapidly becoming easier. Software and hardware are making remarkable advances in ease of use. The future is bright.

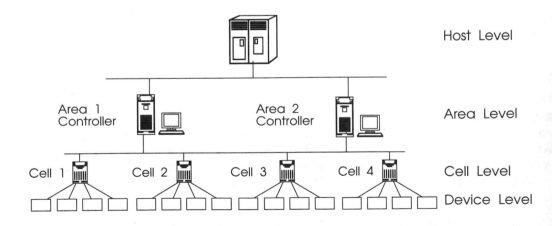

Figure 9-23. Typical host-level control. It is also called the enterprise level.

Chapter 9: Overview of Plant Floor Communication

Questions

1. In what ways are hardware communication hierarchies like human organization?

2. What are the primary characteristics of the device level?

3. What are the main functions of the cell control level?

4. Describe the term <u>serial communications</u>.

5. Describe the term <u>asynchronous</u>.

6. Thoroughly explain the term <u>SCADA</u>.

7. What are the main functions of the area level?

8. What are the main functions of the host level?

9. Describe the term MAP.

10. Describe the term MMS.

11. Why collect so much data from the plant floor?

12. Complete the following table.

	Speed	Length limit	Balanced or Unbalanced	Special Characteristics
RS-232				
RS-422				
RS-423				
RS-485				

Chapter 10

Installation and Troubleshooting

Proper installation is crucial in automated systems. The safety of people and machines is at stake. Troubleshooting and maintenance of systems becomes more crucial as systems become more automatic, more complex, and more expensive. An enterprise cannot afford to have a system down for any length of time. The technician must be able to find and correct problems quickly.

Objectives

Upon completion of this chapter, the student will be able to:

Describe safety considerations that are crucial in troubleshooting and maintaining systems.

Explain such terms as <u>noise</u>, <u>snubbing</u>, <u>suppression</u>, and <u>single point ground</u>.

Explain correct installation techniques and considerations.

Explain proper grounding techniques.

Explain noise reduction techniques.

Explain a typical troubleshooting process.

Installation

Proper installation of A PLC is crucial. The PLC must be wired so that the system is safe for the workers. The system must also be wired so that the devices are protected from over current situations. Proper fusing within the system is important. The PLC must also be protected from the application environment. There are usually dust, coolant, chips, and other contaminants in the air. The proper choice of enclosures can protect the PLC.

Enclosures

PLCs are typically mounted in protective cabinets. The National Electrical Manufacturers Association (NEMA) has developed standards for enclosures (see Figure 10-1). Enclosures are used to protect the control devices from the environment of the application. A cabinet type is chosen based on how severe the environment is in the application. Cabinets typically protect the PLC from airborne contamination. Metal cabinets can also help protect the PLC from electrical noise. Heat is generated by devices. One must make sure that the PLC and other devices to be mounted in the cabinet can perform at the temperatures required. Remember that it will be even hotter in the cabinet than in the application because of the heat generated within the cabinet. Make sure that there is adequate space around the PLC in the enclosure. This will allow air to flow around the PLC. Cooling fans are often unnecessary. Check the specifications for the PLC that will be used to see what temperatures it can operate in.

Protection against	Enclosure Type													
	1	2	3	3r	3s	4	4x	5	6	6p	11	12	12	13
Accidental contact with enclosed equipment	*	*	*	*	*	*	*	*	*	*	*	*	*	*
Falling dirt	*	*				*	*	*	*	*	*	*	*	*
Falling liquids, light splashing		*				*	*		*	*	*	*	*	*
Dust,lint,fibers, (non-cobustible,non-ignitable)						*	*	*	*	*		*	*	*
Windblown dust			*		*	*	*		*	*				
Hose down and splashing water						*	*		*	*				
Oil and coolant seepage												*	*	*
Oil or coolant spraying or splashing														*
Corrosive agents							*			*	*			
Occasional temporary submersion									*	*				
Occasional prolonged submersion										*				

Figure 10-1. Comparison of the features of NEMA cabinets versus enclosure type.

Wiring

Proper wiring of a system involves choosing the appropriate devices and fuses (see Figure 10-2). Normally, three-phase power (typically 480 V) will be used in manufacturing. This will be the electrical supply for the control cabinet.

The three-phase power is connected to the cabinet via a mechanical disconnect. This disconnect is mechanically turned on or off by the use of a lever on the outside of the cabinet. This disconnect should be equipped with a lockout. This means that the technician should be able to put a lock on the lever to prevent anyone from accidentally applying power while it is being worked on. This three-phase power must be fused. Normally, a fusible disconnect is used. This means that the mechanical disconnect has fusing built in. The fusing is to make sure that too much current cannot be drawn. Figure 10-2 shows the fuses installed after the disconnect.

The three-phase power is then connected to a contactor. The contactor is used to turn all power off to the control logic in case of an emergency. The contactor is attached to hardwired emergency circuits in the system. If someone hits an emergency stop switch, the contactor drops out and will not supply power. The hardwired safety should always be used in systems. In addition, the PLC may be able to shut off the contactor.

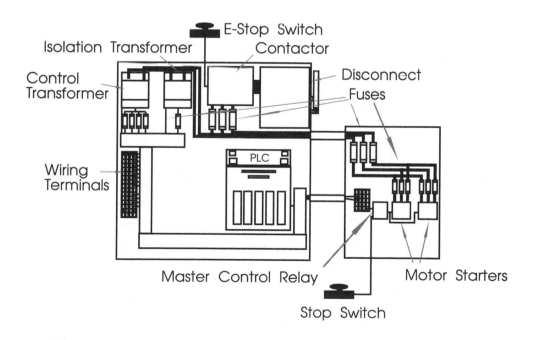

Figure 10-2. Block diagram of a typical wiring scheme.

The three-phase power must then be converted to single-phase for the control logic. Power lines from the fusing are connected to transformers. In the case of Figure 10-2 there are two transformers: an isolation transformer and a control transformer. The isolation transformer is used to clean up the power supply for the PLC.

The control transformer is used to supply other control devices in the cabinet. The lines from the power supplies are then fused to protect the devices they will supply. These individual device circuits should be provided with their own fuses to match the current draw. Dc power is usually required also. A small power supply is typically used to convert the ac to dc.

Motor starters are typically mounted in separate enclosures. This protects the control logic from the noise these devices generate.

Within the cabinet certain wiring conventions are typically used. Red wiring is normally used for control wiring, black wiring is used for three-phase power, blue wire is used for dc, and yellow wire is used to show that the voltage source is separately derived power (outside the cabinet).

Signal wiring should be run separately from 120-V wiring. Signal wiring is typically low voltage or low current and could be affected by being too close to high-voltage wiring. When possible, run the signal wires in separate conduit. Some conduit is internally divided by barriers to isolate signal wiring from higher voltage wiring.

Voltage is supplied to the PLC through wiring terminals. The user can often configure the PLC to accept different voltages (see Figure 10-3). The PLC in this figure can be configured to accept either 220 or 110 V. In this case the shorting bar must be installed if 110 V will be used.

Wiring Guidelines

Here are some guidelines that should be considered when wiring a system. Courtesy of Siemens Industrial Automation, Inc..

Always use the shortest possible cable.

Use a single length of cable between devices. Do not connect pieces of cable to make a longer cable. Avoid sharp bends in wiring.

Avoid placing system and field wiring close to high-energy wiring.

Physically separate field input wiring, output wiring, and other types of wiring.

Separate dc and ac wiring when possible.

A good ground must exist for all components in the system (0.1 ohm or less).

If long return lines to the power supply are needed, do not use the same wire for input and output modules. Separate return lines will minimize the voltage drop on the return lines of the input connections.

Use cable trays for wiring.

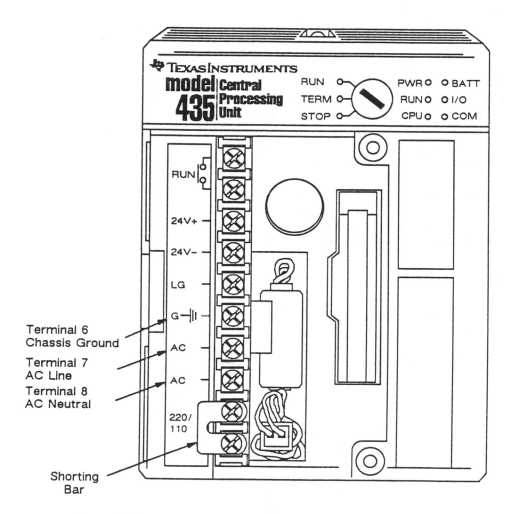

Terminal 6
Chassis Ground

Terminal 7
AC Line

Terminal 8
AC Neutral

Shorting
Bar

Figure 10-3. How to connect power to a PLC CPU. Note the contact labeled "RUN." This can be used as an input to put the PLC into run mode. If the shorting bar is installed, 110 V must be supplied. Courtesy of Siemens Industrial Automation, Inc..

Grounding

Proper grounding is essential for safety and proper operation of a system. Siemens Industrial Automation, Inc. suggests the following guidelines. Use a minimum of No. 12 AWG stranded copper wire for the ground return. Check the appropriate electrical codes to assure compliance with minimum wire sizes, color coding, and general safety practices.

Connect the PLC and components to the subpanel ground bus. The connection should exhibit very low resistance. Connect the subpanel ground bus to a single-point ground, such as a copper bus bar to a good earth ground reference. There must be low impedance between each device and the single-point termination. A rule of thumb would be less than 0.1 ohm dc resistance between device and single-point ground. This can be accomplished by removing the anodized finish and using copper lugs and star washers.

Grounding Guidelines

Grounding braid and green wires should be terminated at both with copper eye lugs to provide good continuity. Lugs should be crimped and soldered. Copper No. 10 bolts should be used for those fasteners that are used to provide an electrical connection to the single-point ground. This applies to device mounting bolts, and braid termination bolts for subpanel and user-supplied single-point grounds. Tapped holes should be used rather than nuts and bolts.

Paint, coatings, or corrosion must be removed from the areas of contact. Use external toothed lock washers to ensure good continuity and low impedance. This practice should be used for all terminations: lug to subpanel, device to lug, device to subpanel, subpanel to conduit, and so on. See Figure 10-4 for examples of ground connections.

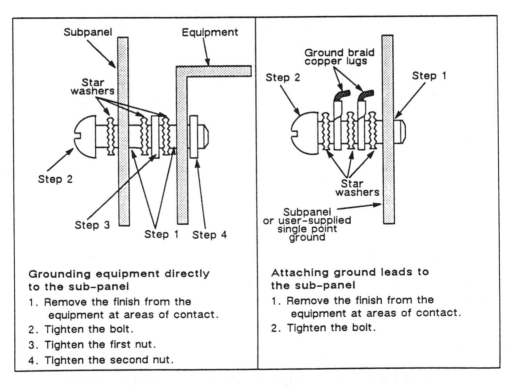

Figure 10-4. Ground connections. Courtesy of Siemens Industrial Automation, Inc..

Handling Electrical Noise

Electrical noise is unwanted electrical interference that affects control equipment. The control devices in use today utilize microprocessors. Microprocessors are constantly fetching data and instructions from memory. Noise can cause the microprocessor to misinterpret an instruction or fetch bad data. Noise can cause minor problems or can cause severe damage to equipment and people.

Noise is caused by a wide variety of manufacturing devices. Devices that switch high voltage and current are the primary sources of noise. These would include large motors and starters, welding equipment, and contactors that are used to turn devices on and off. Noise is not continuous. It can be very difficult to find intermittent noise sources.

Noise can be created by power line disturbances, transmitted noise or ground loops. Power line disturbances are generally caused by devices which have coils. When they are switched off they create a line disturbance. Power line disturbances can normally be overcome through the use of line filters. Surge suppressors such as MOVs or an RC network across the coil can limit the noise. Coil type devices include relays, contactors, starters, clutches/brakes, solenoids, and so on.

Transmitted noise is caused by devices which create radio frequency noise. Transmitted noise is generally caused in high current applications. Welding causes transmitted noise. When contacts that carry high current open, they generate transmitted noise. Application wiring carrying signals can often be disrupted by this type of noise. Imagine wiring carrying sensor information to a control device. In severe cases, false signals can be generated on the signal wiring. This problem can often be overcome by using twisted-pair shielded wiring and connecting the shield to ground.

Transmitted noise can also "leak" into control cabinets. The holes that are put into cabinets for switches and wiring allow transmitted noise to enter the cabinet. The effect can be reduced by properly grounding the cabinet.

Ground loops can also cause noise. These are the noise problems that are often difficult to find. These are quite often intermittent problems. They generally occur when multiple grounds exist. The farther the grounds are apart the more likely the problem. A potential can exist between the power supply earth and the remote earth. This can create unpredictable results, especially in communications.

Proper installation technique can avoid problems with noise.

There are two main ways to deal with noise: suppression and isolation.

Noise Suppression

Suppression attempts to deal with the device that is generating the noise. A very high voltage spike is caused when the current to an inductive load is turned off. This high voltage can cause trouble for the device that controls the output. Some PLC modules include protection circuitry to protect against inductive spikes. A suppression network can be installed to limit the voltage spikes. Figure 10-5 shows how ac and dc loads can be protected against surges. Noise suppression is also called <u>snubbing</u>. Snubbing can be used to suppress the arcing of mechanical contacts caused by turning inductive loads off (see Figure 10-6). Surge suppression should be used on all coils.

An RC or a varistor circuit can be used across an inductive load to suppress noise (1000 ohm, .2 microfarad). Check the installation manual for the PLC that you are using for proper noise suppression.

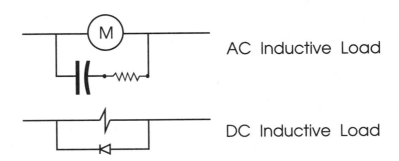

AC Inductive Load

DC Inductive Load

Figure 10-5. Surge suppression methods for ac and dc loads.

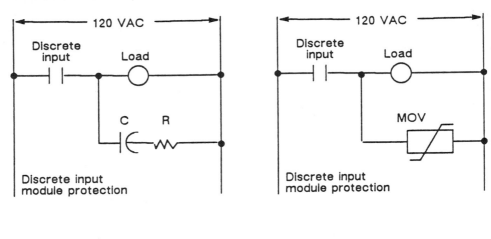

RC type noise snubbing   MOV snubbing

Figure 10-6. This figure shows two examples of noise snubbing. Courtesy of Siemens Industrial Automation, Inc..

Noise Isolation

The other way to deal with noise is isolation. The device or devices that cause trouble are physically separated from the control system. The enclosure also helps separate the control system from noise. In many cases field wiring must be placed in very noisy environments to allow sensors to monitor the process. This presents a problem especially when low voltages are used. Shielded twisted-pair wiring should be used for the control wiring in these cases. The shielding should only be grounded at one end. The shield should be grounded at the single-point ground.

PLC Troubleshooting

The first consideration in troubleshooting and maintaining systems is safety.

When you encounter a problem, remember that less than one-third of all system failures will be due to the PLC. Most of the failures are due to input and output devices (approximately 50 percent).

A few years ago, a technician was killed when he isolated the problem to a defective sensor. He bypassed the sensor and the system restarted with him in it. He was killed by the system he had fixed. You must always be aware of the possible outcomes of changes you make.

Troubleshooting is actually a relatively straightforward process in automated systems. The first step is to think (see Figure 10-7). This may seem rather basic, but many people jump to improper, premature conclusions and waste time finding problems. The first step is to examine the problem logically. Think the problem through using common sense first. This will point to the most logical cause. Troubleshooting is much like 20 questions. Every question should help isolate the problem. In fact, every question should eliminate about half of the potential causes. Remember, a well-planned job is half done.

> *Think logically.*

> *Ask yourself questions to isolate the problem.*

> *Test your theory.*

Next use the resources you have available to check your theory. Often the error-checking present on the PLC modules is sufficient. The LEDs on PLC CPUs and modules can provide immediate feedback on what is happening. Many PLCs have LEDs to indicate blown fuses and many other problems. Check these indicators first.

The usual problem is that an output is not turning on when it should. There are several possible causes. The output device could be defective. The PLC output that turns it on could be defective. One of the inputs that allow, or cause, the output to turn on could be defective. The sensor or the PLC input that it is attached to could be defective. The ladder logic could even be faulty. It is possible that a ladder can be written that performs perfectly the vast majority of the time but fails under certain conditions. Again the module I/O LEDs provide the best source of answers. If the PLC module output LED is on for that output, the problem is probably not the inputs to the PLC. The device is defective, the wiring is defective, or the PLC output is defective.

The next step is then to isolate the problem further. A multimeter is invaluable at this point. If the PLC output is off, a meter reading should show the full voltage with which the device is turned on. If the output is used to supply 115 V to a motor starter, the meter should show the full 115 volts between the output terminal and common on the PLC module (see Figure 10-8).

Output Troubleshooting Chart

Output Condition	Output LED	Status in Ladder	Probable Problem
on	on	true —()—	None
off	off	true —()—	Bad fuse or bad output module
off	off	false —O—	None
off	on	true —()—	Wiring to output device or bad output device

Figure 10-7. How to isolate a PLC problem by comparing the states of inputs/outputs, indicators, and the ladder status.

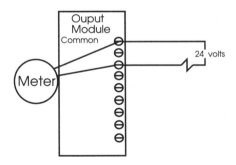

Figure 10-8. How to test an output. This will help isolate the problem for the technician.

If the PLC output is on, the meter should read zero volts because the output acts as a switch. If we measure the voltage across a switch, we should read zero volts. If the switch is open, we should read the full voltage. If we read no voltage in either case, the wiring and power supply should be checked. If the wiring and power supply are good, the device is the problem. Depending on the device, there may be fuses or overload protection present.

Now we will step back in our troubleshooting procedure and pretend that the output LED was not turning on. Now we must check the input side. If the input LED is on, we can assume that the sensor, or other input device, is operational (see Figures 10-9 and 10-10). Next we must see if the PLC CPU really sees the input as true. At this point a monitor of some type is required. A hand-held programmer or a computer is often used. The ladder is then monitored under operation. Note: Many PLCs allow the outputs to be disabled for troubleshooting. This is the safe way to proceed. Check the ladder to see if the contact is closing. This tells us whether or not the CPU is actually seeing the input. If it is seeing the input as false, the problem is probably a defective PLC module input.

Input Troubleshooting Guide

Actual input condition	Input module LED status	Ladder Status		Probable problem
off	off	false ─┤ ├─	true ─┤/├─	None
off	on	true ─┤■├─	false ─┤/├─	Short in the input device or wiring or a bad input module
on	off	false ─┤ ├─	true ─┤/├─	wiring/power to I/O module or I/O module
on	on	false ─┤ ├─	true ─┤/├─	I/O module
on	on	true ─┤■├─	false ─┤/├─	None

Figure 10-9. Chart that can be used for troubleshooting inputs.

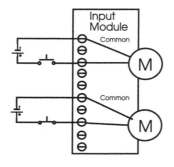

Figure 10-10. This figure shows how to isolate input problems.

Potential Ladder Diagram Problems

Some PLC manufacturers allow output coils to be used more than once in ladders. This means that there can be multiple conditions that can control the same output coil. I have seen technicians who were sure that they had a defective output module because when they monitored the ladder the output coil was on but the actual output LED was off. In this case the technician had inadvertently programmed the same output coil twice with different input logic. The rung he was monitoring was true, but one farther down in the ladder was false, so the PLC kept the output off.

The other potential problem with ladder diagramming is that the problem can be intermittent. Timing is crucial in ladder diagramming. Devices are often exchanging signals to signify that an action has occurred. For example, a robot finishes its cycle and sends a digital signal to the PLC to let it know. The programmer of the robot must be aware of the timing considerations. If the robot programmer just turns it on for one step the output may only be on for a few milliseconds. The PLC may be able to "catch" the signal every time, or it may occasionally miss it because of the length of the scan time of the ladder diagram. This can be a hard problem to find, because it only happens occasionally. The better way to program is to handshake instead of relying on the PLC seeing a periodic input. The programmer should have the robot output stay on until it is acknowledged by the PLC. The robot turns on the output and waits for an input from the PLC to assure that the PLC saw the output from the robot.

Summary

The installation of a control system must be carefully planned. People and devices must be protected. Hardwired switching should be provided to drop all power to the system. Lockouts should be provided to assure safety while maintenance is being performed. Proper fusing to assure protection of individual devices is a must. The cabinet must be carefully selected to meet the needs of the application environment. Control and power wiring should be separated to reduce noise. Proper grounding procedures must be followed to ensure safety.

Troubleshooting must be done in a logical way. Think the problem through. Ask questions of yourself that will help isolate the potential problems. Above all, apply safe work habits while working on systems.

Questions

1. What is a NEMA enclosure?

2. Why should enclosures be used?

3. Describe how an enclosure is chosen.

4. What is a fusible disconnect?

5. What is a contactor?

6. What is the purpose of an isolation transformer?

7. What is the major cause of failure in systems?

8. Describe a logical process for troubleshooting.

9. Describe proper grounding techniques.

10. Draw a block diagram of a typical control cabinet.

11. Describe at least three precautions that should be taken to help reduce the problem of noise in a control system.

12. A technician has been asked to troubleshoot a system. The output device is not turning on for some reason. The output LED is working as it should. The technician turns the output on and places a meter over the PLC output. 115 V is read. The device is not running. What is the most likely problem? How might it be fixed?

13. A technician has been asked to troubleshoot a system. The PLC does not seem to be receiving an input because the output it controls is not turning on. The technician notices that the input LED is not turning on. The technician notices that the output indicator LED on the sensor seems to be working. A meter is placed across the input with the sensor on. 24 V is sensed, but the input LED is off. What is the most likely problem?

14. A technician is asked to troubleshoot a system. An input seems to be defective. The technician notices that the input LED is never on. A meter is placed across the PLC input and zero volts is read. The technician removes and tests the sensor. (The LED on the sensor comes on when the sensor is activated.) The sensor is fine. What is the most likely problem?

15. A technician is asked to troubleshoot a system. An output device is not working. The technician notices that the output LED seems to be working fine. A meter is placed over the PLC output with the output on. Zero volts is read. Describe what the technician should do to find the problem.

Appendix A

In this appendix we examine some of the commonly used instructions for the Allen Bradley PLC-5 family.

The memory organization for data files is shown in Figure A-1. The chart shows the maximum number of elements of each type, the file type, the number that is used to designate the type of file, and the number of words per element. The programmer designates the type of file according to the type of data that will be stored in it.

Max # of Elements	File type	File #	Words/ element
32*	Output Image	0	1
32*	Input Image	1	1
32	Status	2	1
1000	Bit	3	1 (16 bits)
1000	Timer	4	3
1000	Counter	5	3
1000	Control	6	3
1000	Integer	7	1
1000	Floating point	8	2
1000	File type as needed	9-999	
* Up to 64 for the PLC-5/25 processor			

Figure A-1. Memory organization for a PLC-5.

Most files use one word elements. For these each bit can be addressed individually. The range of values that can be stored in one word of memory is -32,768 to +32,767. Floating-point files use two words of memory. The floating point element must be addressed as a whole. Individual bits cannot be accessed. Timer and counter files use three words of memory. The words (or elements) are used to store the status bits, .PRE (preset) values, and .ACC (accumulated) values. These are covered in more detail later.

Logical Addressing

The Allen Bradley system of addressing allows you to address bits, elements (words), or data files. The format varies slightly depending on the type of file (see Figure A-2). There are occasions when the status of a particular bit must be accessed. There are also occasions when a whole word must be accessed.

#Xf:e.s/b

 # File-address identifier

X File Type

B=bit	C=Counter
F=Floating point	I=Input
N=Integer	O=Output
R=Control	S=Status
T=Timer	A=ASCII-For display only
D=BCD	

File Number

0=Output	1=Input
2=Status	3=Bit
4=Timer	5=Counter
6=Control	7=Integer
8=Floating point	9-999 for additional files

: Separates file and element numbers

e Element number

0-37 Octal for I/O files
0-31 Decimal for the status file
0-999 Decimal for all other file types

s Use only with counter, timer, and control files

.PRE(preset) .ACC(accumulated)	.LEN(length) .POS(position)

/ Delimiter separates bit numbers from element or subelement numbers

b Bit number

0-17 Octal for I/O files	0-15 for all other file types

Figure A-2. This figure shows the addressing scheme.

I/O addressing

I/O addressing varies from the general format. Figure A-3 shows the method in which I/Os are addressed. I/O are addressed by the programmer specifying an "I" or an "O" for input or output, the rack number, the I/O group number, and the terminal (bit) number.

$$O:rg/00-17 \qquad\qquad I:rg/00-17$$

I = Input
O = Output
r = Assigned rack number
g = I/O group number
00-17 = bit (terminal) number

Figure A-3. How I/O addressing is done.

Status files are addressed by the programmer specifying the element number and the bit number (see Figure A-4). Bits or elements (words) can be addressed. Note that individual bits can be addressed and used in ladder logic for most file types (see Figure A-5).

$$S:e/b$$

S = Status
e = Element number (0-31)
b = Bit number (0-15)

Figure A-4. How status files are addressed.

Bit Addressing Examples

B3:15/0   This is binary file 3, element 15, bit 0

I:02/7   This is input 7 of I/O group 2, rack 0.

C5:6.DN   This is counter file 5, element 6, bit 13. (bit 13 is the done bit)

N7:3/4   This is the format for the bit address in an integer file. It is integer file 7, element 3, bit number 4.

Figure A-5. Bit addressing examples.

It is often desirable to access a whole element (word) rather than a single bit. For example, if the programmer needed to access the accumulated count of a counter or the accumulated time of a timer to make a logic decision, a whole element would be accessed. Elements (words) can be accessed by specifying the file type and element to be addressed. Figure A-6 shows examples of element addressing.

Element Address Format

S:3   This is the format for addressing element 3 in a status file.

N7:3   This is the format for addressing element 3 in integer file number 7.

F8:53   This is the format for addressing element 53 in floating point file number 8.

Figure A-6. Examples of element addressing.

Examine-On

The examine-on instruction (XIC) is the normally open instruction. If a real-world input device is on, this type of instruction is true and passes power (see Figure A-7).

If the input bit from the input image table associated with this instruction is a one, the instruction is true. If the bit in the input image table is a zero, the instruction associated with this particular input bit is false.

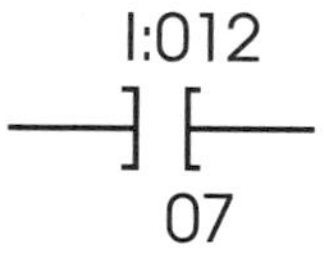

I:012

―] [―

07

Figure A-7. Examine-on instruction (XIC). If the CPU sees an on condition at bit I:012/07, this instruction is true. The numbering of the input instruction is as follows: This is rack 1, I/O group 2, real-world input 7 of the input module.

Examine-Off

The examine-off instruction (XIO) is also called a normally closed instruction. This instruction responds in the opposite fashion to the normally open instruction. If the bit associated with this instruction is a zero (off), the instruction is true and passes power. If the bit associated with the instruction is a one (true), the instruction is false and does not allow power flow.

Figure A-8 shows an examine-off instruction (XIO). The input is number 7 of rack 1, I/O group 2. If the bit associated with I:012/07 is true (one) the instruction is false (open) and does not allow power flow. If bit I:012/07 is false (zero) the instruction is true (closed) and allows flow.

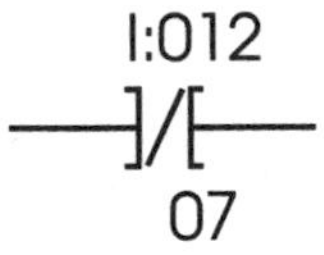

$$I:012$$
$$\dashv/\vdash$$
$$07$$

Figure A-8. Examine-off instruction (normally closed).

Output Energize

The output energize instruction (OTE) is the normal output instruction. The OTE instruction sets a bit in memory. If the logic in its rung is true, the output bit will be set to a 1. If the logic of its rung is false, the output bit is reset to a zero. Figure A-9 shows an output energize (OTE) instruction. This particular example is real-world output 1 of I/O group 3, of rack 1. If the logic of the rung leading to this output instruction is true, output bit O:013/01 will be set to a 1 (true). If the rung is false, the output bit would be set to a zero (false).

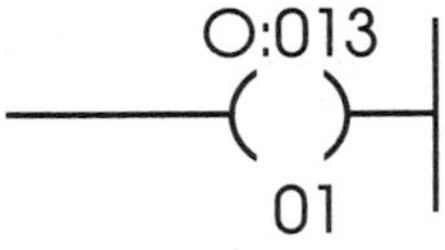

$$O:013$$
$$-(\ )-$$
$$01$$

Figure A-9. Output energize instruction (OTE).

Output Latch

The output latch instruction (OTL) is a retentive instruction. If this input is turned on, it will stay on even if its input conditions become false. A retentive output can only be turned off by an unlatch instruction. Figure A-10 shows an output latch instruction (OTL). In this case if the rung conditions for this output coil are true, output bit O:013/01 will be set to a 1. It will remain a 1 even if the rung becomes false. Output 1 of I/O group 3, rack 1 will be "latched" on. Note that if the OTL is retentive, if the processor loses power the actual output turns off, but when power is restored the output is retentive and will turn on. This is also true in the case of switching from run to program mode. The actual output turns off, but the bit state of 1 is retained in memory. When the processor is switched to run again, retentive outputs will turn on again regardless of the rung conditions. Retentive instructions can help or hurt the programmer. Be very careful from a safety standpoint when using retentive instructions.

$$O:013$$
$$-(L)-$$
$$01$$

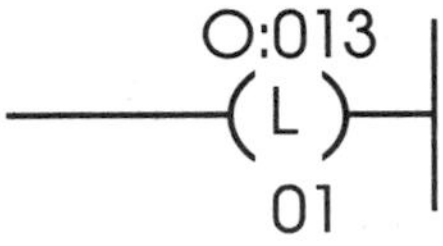

Figure A-10. Output latch instruction (OTL).

Output Unlatch

The output unlatch instruction (OTU) is used to "unlatch" (change the state of) retentive output instructions. It is the only way to turn an output latch instruction (OTL) off.

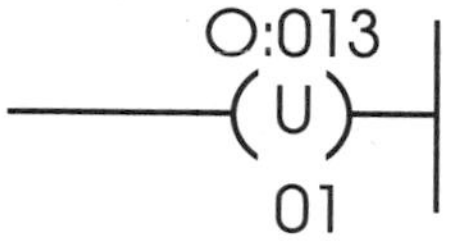

$$O:013$$
$$-(U)-|$$
$$01$$

Figure A-11. Output unlatch instruction (OTU).

Figure A-11 shows an output latch instruction (OTU). In this case if the bit associated with real-world input 1 of I/O group 3 of rack 1 is true, the instruction is true. If this instruction is true, it unlatches the retentive output of the same number. In this case output bit O:013/01 would be reset to a 1 and the actual output would be off.

Immediate Input

The immediate input instruction (IIN) is an instruction that is used to acquire the present state of one word of inputs. Normally, the CPU would have to finish all evaluation of the ladder logic and then update the output and input image tables. In this case when the CPU encounters this instruction during ladder evaluation, it drops what it is doing for a short time and updates one word (16 inputs) of the input image table. This means that it gets the real-time states of the actual inputs at that time and puts them in that word of the input image table. The CPU then returns to evaluating the logic using the new states it acquired. This is used only when time is a crucial factor. Normally, the few milliseconds a scan takes is fast enough for anything we do. There are cases where the scan time takes too long for some I/O updates. Motion control is one case. Updates on speed and position may be required every few milliseconds or less. We cannot safely wait for the scan to finish to update the I/O. In these cases immediate instructions are used.

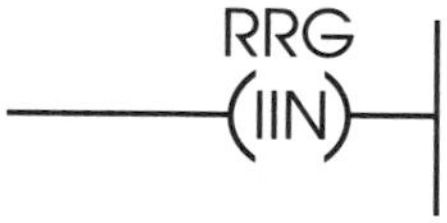

$$RRG$$
$$-(IIN)-|$$

Figure A-12. Instruction numbering. This is an immediate input instruction (IIN).

Figure A-12 shows an immediate input instruction (IIN) example format. The RRG in the numbering represents what the user must enter. The RR is equal to the I/O rack number of the desired input states to be updated. The G represents the I/O group number of the word of input memory to be updated. Figure A-13 shows an immediate input instruction. When the rung conditions for this instruction are true, the instruction will be true. When the CPU encounters the instruction during evaluation it will immediately suspend what it is doing (evaluating) and will update the input image word associated with I/O rack, group 1.

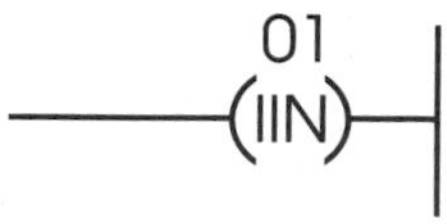

$$01$$
$$-(IIN)-|$$

Figure A-13. Immediate input instruction (IIN).

Immediate Output

The immediate output instruction (IOT) is used to update output states immediately. In some applications the ladder scan time is longer than the needed update time for certain outputs. For example, it might cause a safety problem if an output were not turned on or off before an entire scan was complete. In these cases, or when performance requires immediate response, immediate outputs (IOTs) are used.

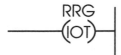

Figure A-14. Numbering format for a immediate output instruction (IOT).

Figure A-14 shows an example of the output format for an IOT. The RRG stands for the information that the programmer would enter. The RR stands for the rack number. The G stands for the group number. Figure A-15 shows the use of an immediate output instruction. In this case if the rung is true, the CPU will immediately update the output word associated with I/O rack 0, group 1. The CPU will continue evaluating the ladder logic.

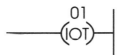

Figure A-15. Immediate input instruction.

Timer On Delay

The timer on delay instruction (TON) is used to turn an output on after a timer has been on for a preset time interval. The timer on delay (TON) begins accumulating time when the rung becomes true and continues until one of the following conditions is met: the accumulated value is equal to the preset value; the rung goes false; a reset timer instruction resets the timer; or the associated SFC step becomes inactive. Figure A-16 shows numbering system for timers. The T stands for timer, the X would be the file number of the timer, and the Y is the actual timer number.

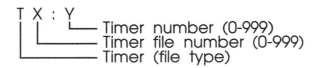

Figure A-16. Format for a timer.

Condition:	Result:
If the rung is true	.EN bit remains set (1)
	.TT bit remains set (1)
	.ACC value is reset to zero
If the rung is false	.EN bit is reset
	.TT bit is reset
	.DN bit is reset
	.ACC value is cleared and begins counting up

Figure A-17. Use of special timer bits.

Status Bit Use

The .sb in the first block in Figure A-17 represents how the timer status bits can be used in ladder logic. There are several bits available for use. The timer enable bit (.EN) is set when the rung goes true. It stays set until the rung goes false, or a reset instruction resets the timer, or the associated SFC step goes false. The .EN bit indicates that the timer is enabled. The .EN bit from any timer can be used for logic. For example, T4:0.EN could be used as the number for a contact in a ladder.

The timer timing bit (.TT) can also be used. The .TT bit is set when the rung goes true. It remains true until the rung goes false, or the .DN bit is set (accumulated value equals preset value), or the associated SFC step becomes inactive. For example, T4:0.TT could be used as a contact in a ladder.

The timer done bit (.DN) is set until the accumulated value is equal to the preset value and the .DN remains set until the rung goes false, a reset instruction resets the timer, or the associated SFC step becomes inactive. When the .DN bit is set, it is an indication that the timing operation is complete. For example, T4:0.DN could be used as a contact in the ladder. The preset (.PRE) is also available to the ladder. For example, T4:0.PRE would access the preset value of T4:0.

Accumulated Value Use

The accumulated value can also be used by the programmer. The accumulated value (.ACC) is acquired in the same manner as the status bits and preset. For example, T4:0.ACC would access the accumulated value of timer T4:0.

Time Bases

Two time bases are available; 1-second intervals, or 0.01 second intervals (see Figure A-18). The potential time ranges are also shown. If a longer time is needed, timers can be cascaded (see Chapter 4). Figure A-19 shows how timers are handled in memory. Three bits are used in the first storage location for this timer to store the present status of the timers bits (EN, .TT, and .DN). The preset value (PRE) is stored in the second 16 bits of this timer storage. The third 16 bits hold the accumulated value of the timer.

Time Base	Potential Time Range
1 second	To 32,767 time-base intervals (up to 9.1 Hours)
.01 seconds (10 ms)	To 32,767 time-base intervals (up to 5.5 minutes)

Figure A-18. Time bases available.

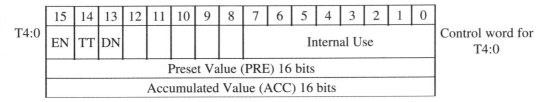

Figure A-19. Use of control words for timers.

Figure A-20 shows an example of the use of a TON timer in a ladder. When input I:012/10 is true, the timer begins to increment the accumulated value of TON timer 4:0 in 1-second intervals. The timer timing bit (TT) for timer 4:0 is used in the second rung to turn on output O:013/01, while the timer is timing (.ACC < .PRE). The timer done bit (DN) of timer 4:0 is used in rung 3 to turn on output O:013/02 when the timer is done timing (.ACC = .PRE). The preset for this timer is 180, which means that the timer will have to accumulate 180 1-second intervals to time out. Note that this is not a retentive timer. If input I:012/10 goes low before 180 is reached, the accumulated value is reset to zero.

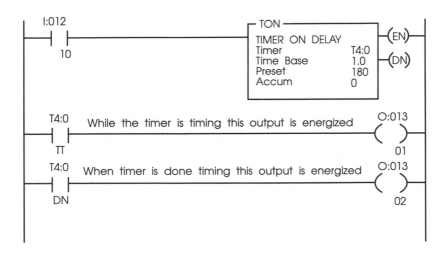

Figure A-20. Use of a TON timer in a ladder logic program.

Timer Off Delay

The timer off delay instruction is used (TOF) to turn an output on or off after the rung has been off for a desired time. The TOF instruction starts to accumulate time when the rung becomes false. It will continue to accumulate time until the accumulated value equals the preset value, or the rung becomes true, or a reset timer instruction resets the timer, or the associated SFC becomes inactive. The timer enable bit (EN bit 15) is set when the rung becomes true. It is reset when the rung become false, or a reset instruction resets the timer, or the associated SFC step becomes inactive. The timer timing bit (TT bit 14) is set when the rung becomes false and ACC < PRE. The TT bit is reset when the rung becomes false, or the DN bit is reset (.ACC=.PRE), or a reset instruction resets the timer, or the associated SFC step becomes false.

The done bit (DN bit 13) is reset when the accumulated value (ACC) is equal to the preset (PRE) value. The DN bit is set when the rung becomes true, or a reset instruction resets the timer, or the associated SFC step becomes inactive.

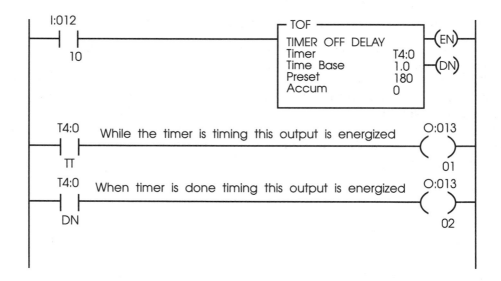

Figure A-21. Use of a TOF timer in a ladder logic diagram.

Figure A-21 shows the use of a TOF timer in a ladder diagram. Input I:012/10 is used to enable the timer. When input I:012/10 is false the accumulated value is incremented as long as the input stays false and ACC < = PRE. The timer enable bit for timer T4:0 (T:4.TT) is used to turn output O:013/01 on while the timer is timing. The done bit (DN) for timer 4:0 (T4:0.DN) is used to turn on output O:013/02 when the timer has completed the timing (ACC = PRE).

Retentive Timer On

The retentive timer on instruction (RTO) is used to turn an output on after a set time period (see Figure A-22). The RTO timer is an accumulating timer. It retains the accumulated value even if the rung goes false. The only way to zero the accumulated value is to use a reset instruction in another rung with the same number as the RTO you wish to reset. The RTO retains the accumulated count even if power is lost, or you switch modes, or the rung becomes false, or the associated SFC becomes inactive. Remember, the only way to zero the accumulated value is to use a reset instruction.

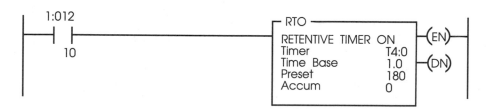

Figure A-22. Use of a RTO timer.

The status bits can be used as contacts in a ladder diagram. The timer enable bit (.EN) is set when the rung becomes true. When the .EN bit is a 1 it indicates that the timer is timing. It remains set until the rung becomes false or a reset instruction zeros the accumulated value.

The timer timing bit (.TT) is set when the rung becomes true and remains set until the accumulated value equals the preset value or a reset instruction resets the timer. When the .TT bit is a one it indicates that the timer is timing. The .TT bit is reset when the rung becomes false or when the done bit (.DN) is set. The timer done bit (.DN) is set when the timers accumulated value is equal to the preset value. When the .DN bit is set it indicates that the timing is complete. The .DN is reset with the reset instruction.

Counters

Counters are programmed almost exactly like timers. There is a counter number, a preset, and an accumulated value. The counter is numbered like the timer except that it begins with a "C." The next number is a file number between 0 and 999. The third value is the counter number, also 0 through 999. (See Figure A-23.)

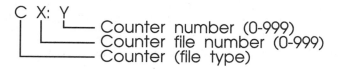

Figure A-23. How counters are addressed.

Your ladder diagram can access counter status bits, presets, and accumulated values (see Figure A-24). The .sb stands for status bit. You may use the .CU, .CD, .DN, .OV, or .UN bit for logic. These will each be covered a little later. You can also use the preset (.PRE) and the accumulated count (.ACC).

Counter values are stored in three 16 bit words of memory. The first eight bits of the first word are only for internal use of the CPU. The most significant bits of the first word are used to store the status of certain bits associated with the counter (see Figure A-25). The count-up enable

(.CU bit 15) bit is used to indicate that the counter is enabled. The .CU bit is reset when the rung becomes false or when it is reset by a RES instruction. The count-up done (.DN bit 13) bit when high indicates that the accumulated count has reached the preset value. It remains set even when the accumulated value (.ACC) exceeds the preset value (.PRE). The .DN bit is reset by a reset (RES) instruction.

Status Bit	Preset	Accumulated Value
CX:Y.sb	CX:Y.PRE	CX:Y.ACC

Figure A-24. How counter bits and values can be accessed in a ladder diagram.

The count-up overflow (.OV bit 12) bit is set by the CPU to show that the count has exceeded the upper limit of +32,767. When this happens the counter accumulated value "wraps around" to -32,768 and begins to count up from there, back toward zero. (This has to do with the way computers store negative numbers. The preset value and accumulated value is stored as a two's complement number.) The .OV bit can be reset with a reset (RES) instruction.

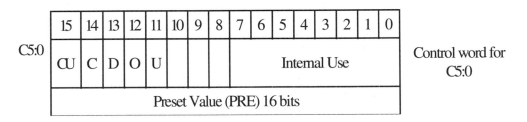

Figure A-25. How counter values and status bits are stored in memory.

Figure A-26 shows the use of a count-up counter (CTU) in a ladder diagram. Each time input I:012/10 makes a low-to-high transition the counter accumulated value is incremented by one. The done bit of counter 5:0 (C5:0.DN) is being used to turn output I:013/01 on when the accumulated value is equal to the preset value (.ACC = .PRE). The overflow bit of counter 5:0 (C5:0.OV) is being used to set output I:013/02 on if the count ever reaches +32,767. The last rung uses input I:017/12 to reset counter 5's accumulated value to zero.

Figure A-27 shows the use of a count-down counter (CTD) in a ladder diagram. Each time input I:012/10 makes a low-to-high transition the counter accumulated value is decremented by one. The done bit of counter 5:0 (C5:0.DN) is being used to turn output I:013/01 on when the accumulated value is equal to or exceeds the preset value (.ACC = .PRE). The accumulated value of counters is retentive. They are retained until a reset instruction is used. The underflow bit of counter 5:0 (C5:0.OV) is being used to set output I:013/02 on if the count ever underflows -32,768. Note the use of the reset instruction to reset the accumulated value of the counter to zero. The last rung uses input I:017/12 to reset counter 5's accumulated value to zero (see Figure A-28).

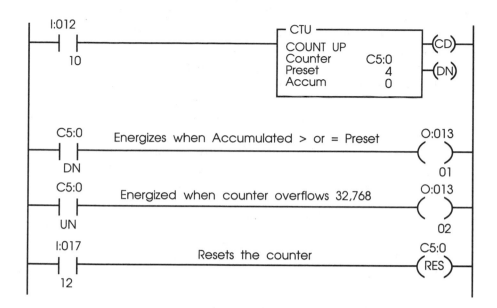

Figure A-26 Use of a count-up counter (CTU) in a ladder diagram.

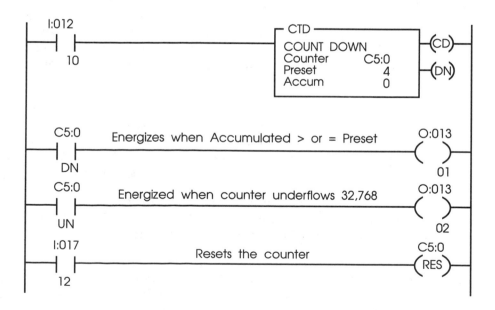

Figure A-27. Use of a count-down counter (CTD).

Reset instruction use	The CPU resets:
Timer (Do not use a reset instruction for a TOF timer.)	.ACC Value .EN bit .TT bit .DN bit
Counter	.ACC Value .EN bit .OV or .UN bit .DN bit

Figure A-28. How the use of a reset (RES) instruction affects timer and counter values and bits.

Master Control Reset Instructions

A master control reset instruction (MCR) is used to control sections of programs. They are used to create control zones in a program. When the MCR is low, all nonretentive outputs in the zone are turned off. When the MCR is high, the ladder logic of the zone controls the outputs.

The first MCR signals the start of the zone. The second MCR shows the end of the controlled zone. The use of these MCRs should not be confused with the use of real-world hardwired master control relays to drop all power to a cell in the case of an emergency. The ladder logic MCR does not meet that requirement.

The use of zones allows the programmer to enable or disable entire sections of ladder logic. This can simplify the programmer's task in some cases. If the zone continues to the end of the ladder diagram, a MCR is not required at the end. You cannot nest MCR zones inside each other. Each MCR zone must be separate and complete. Do not overlap them or nest them. Never use a jump instruction to enter a MCR zone. If the zone is false, jumping into it activates the zone.

You must be very careful about the use of timers and counters in MCR zones. They will not increment when the zone is disabled. Critical events should be programmed outside zones when necessary. Figure A-29 shows the use of a MCR zone in a ladder diagram.

The first MCR designates the start of the zone. The second unconditional MCR designates the end of the zone. When input I:012/10 is true, the MCR zone is enabled. The logic of the rungs between the MCRs is active. If input I:012/05 is true, output O:013/01 will be turned on. If input I:012/06 is true in the third rung, output O:13/02 will be turned on. The MCR zone will remain active until contact I:012/05 becomes false. When the MCR zone is not enabled, the CPU resets all nonretentive outputs. It should be remembered that the rest of the ladder diagram (if there is more) is evaluated whether or not the MCR zone is enabled or disabled.

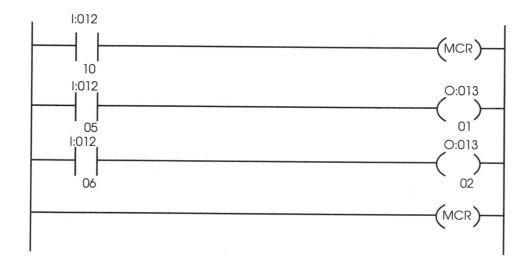

Figure A-29. Use of MCR instructions to create a MCR control zone.

Arithmetic Instructions

There are a multitude of arithmetic instructions available in the AB-5 and SLK-500 controllers. A few of them will be covered in this section.

Compare

The compare instruction (CMP) is used to compare two values or expressions to check on their relationship.

The compare instruction (CMP) is an input instruction that uses mathematical operators to perform comparisons of values or expressions (see Figure A-30). When the comparison is true, the rung is true. The compare instruction execution time is longer than other instructions, such as equal to (EQU). The compare instruction should be used when the programmer needs to perform some mathematical manipulation on the values before comparing them.

The expression portion of the instruction defines what operations the programmer wants to perform. The programmer can use constants or addresses. An expression can be up to 80 characters long. Program constants can be integer or floating-point. If octal numbers are entered a &O must precede the number so that the CPU knows that it is an octal number. If a hexadecimal number is used, it must be preceded by &H. If binary numbers are used, they must be preceded by &B.

The use of a compare instruction (CMP) is shown in Figure A-31. The compare (CMP) instruction tells the CPU to compare the sum of the values in N7:0 and N:7:1 to the difference of the values in N7:2 and N7:3. If the sum of the first two values is greater than the result of the subtraction of the other two values, the rung will be true and output O:013/01 will be turned on.

Operator	Description	Example Operation
=	equal	if a=b, then...
<>	not equal	if a<>b, then...
<	less than	if a<b, then...
<=	less than or equal	if a<=b, then
>	greater than	if a>b, then...
>=	greater than or equal	if a>=b, then...

Figure A-30. Operators that can be used with a compare instruction.

```
 ┌─ CMP ──────────────────┐                                    O:013
─┤ │                      │ ──────────────────────────────────( )──
 │ COMPARE                │                                     01
 │ Expression             │
 │ (N7:0+N7:1) > (N7:2-N7:3) │
 └────────────────────────┘
```

Figure A-31. Use of a compare instruction.

Equal To

The equal to instruction (EQU) is used to test if two values are equal. The values tested can be actual values or addresses that contain values. An example is shown in Figure A-32. Source A is compared to source B to test to see if they are equal. If the value in N:7:5 is equal to the value in N7:10, the rung will be true and output O:013/01 will be turned on.

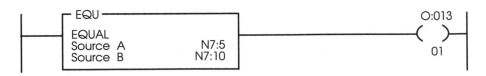

Figure A-32. An equal to instruction (EQU).

Greater Than or Equal to

The greater than or equal to instruction (GEQ) is used to test two sources to determine whether or not source A is greater than or equal to the second source. The use of a GEQ instruction is shown in Figure A-33. If the value of source A (N7:5) is greater than or equal to source B (N7:10), output O:013/01 will be turned on.

Figure A-33. Use of a greater than or equal to (GEQ) instruction.

Greater Than

The greater than instruction (GRT) is used to see if a value from one source is greater than the value from a second source. An example of the instruction is shown in Figure A-34. If the value of source A (N7:5) is greater than the value of source B (N7:10), output O:013/01 will be set (turned on).

Figure A-34. Use of a greater than (GRT) instruction

Less Than

The less than instruction (LES) is used to see if a value from one source is less than the value from a second source. An example of the instruction is shown in Figure A-35. If the value of source A (N7:5) is less than the value of source B (N7:10), output O:013/01 will be set (turned on).

Figure A-35. Use of a less than (LES) instruction.

Limit

The limit instruction (LIM) is used to test a value to see if it falls in a specified range of values. The instruction is true when the tested value is within the limits. This could be used, for example, to see if the temperature of an oven was within the desired temperature range. In this case the instruction would be testing to see if an analog value (a number in memory representing the actual analog temperature) was within certain desired limits.

The programmer must provide three pieces of data to the LIM instruction when programming. The programmer must provide a low limit. The low limit can be a constant or an address that contains the desired value. The address will contain an integer or floating-point value (16 bits).

The programmer must also provide a test value. This is a constant or the address of a value that is to be tested. If the test value is within the range specified, the rung will be true. The third value the programmer must provide is the high limit. The high limit can be a constant or the address of a value.

Figure A-36 shows the use of a limit (LIM) instruction. If the value in N7:15 is greater than or equal to the lower limit value (N7:10) and less than or equal to the high limit (N7:20), the rung will be true and output O:013/01 will be turned on.

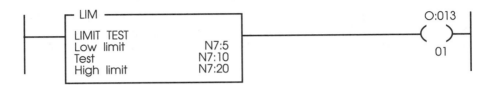

Figure A-36. Use of a limit test (LIM) instruction.

Mask Compare Equal To

The mask compare equal to instruction is used to compare two values to check for equality (see Figure A-37). The difference between this instruction and a regular equal to (EQU) instruction, is that the MEQ instruction permits the masking of bits so that they are not considered in the comparison. This can permit the programmer to consider parts of words when making a ladder decision. Bits that do not matter can be masked. The mask value is what determines whether or not a particular bit is compared. If the bit in the mask is a 1, the bit is used in the comparison. If the bit of the mask is a 0, it is not used in the comparison.

Source	0	1	0	1	0	1	0	1	0	1	0	1	1	1	1	1
Mask	1	1	1	1	1	1	1	1	1	1	1	1	0	0	0	0
Reference	0	1	0	1	0	1	0	1	0	1	0	1	x	x	x	x

Result — THE INSTRUCTION IS TRUE. REMEMBER THAT REFERENCE BITS xxxx ARE NOT COMPARED BECAUSE THE FIRST FOUR BITS IN THE MASK ARE A ZERO AND ARE NOT COMPARED.

Figure A-37. MEQ comparison. The instruction would be true in this case.

The actual use of a MEQ instruction is shown in Figure A-38. The source (N7:5) is compared to the compare value N7:10. Any bits that are a 1 in the mask value (N7:6) will be compared; any bits that are a zero will not be compared. If the compared bits are all the same, the instruction is true. All values used in a MEQ instruction must be 16 bits. The values used are unchanged by the instruction. If the user wishes to change the value of the mask, an address should be used.

The value of the address can then be modified by other instructions. A hex value (constant) can also be used for a mask value. When entering a hex value that begins with a letter such as D170, you must enter the value with a leading zero (0D170).

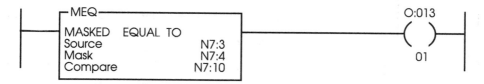

Figure A-38. Use of a masked equal to (MEQ) instruction.

Not Equal To

The not equal to instruction (NEQ) is used to test two values for inequality. The values tested can be constants or addresses that contain values. An example is shown in Figure A-39. If source A (N7:5) is not equal to source B (N7:10), the instruction is true and output O:013/01 is turned on.

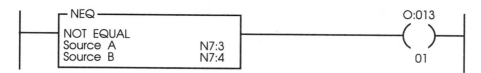

Figure A-39. Use of a not equal to (NEQ) instruction.

Compute Instructions

Compute instructions can be used for all common mathematical operations. They can be used to copy values from one source to another, clear values, and to perform various math operations such as arithmetic, bitwise, and conversion of values from one number system to another. A few of the more common instructions are covered next.

Compute

The compute instruction (CPT) can be used to convert between number systems, manipulate numbers, or perform trigonometric functions.

Remember that floating-point numbers are 32-bit numbers and integer numbers are 16 bits long. The instruction converts numbers from the expression to the type required by the destination address. If the conversion requires a 32-bit number to be changed to a 16-bit number and the result is too large for a 16-bit number, the CPU will set an overflow bit in S2:01 and will also set a minor fault bit (bit 14). The resulting converted number could cause problems. You should monitor these status bits in your program.

Operations are performed in a prescribed order. Operations of equal order are performed left to right. Figure A-40 shows the order in which operations are performed. The programmer can override precedence order by using parentheses.

Order	Operation	Description
1	**	exponential (x to the power of y)
2	-	negate
	NOT	bitwise complement
3	*	multipy
	|	divide
4	+	add
	-	subtract
5	AND	bitwise AND
6	XOR	bitwise exclusive OR
7	OR	bitwise OR

Figure A-40. Precedence (order) in which math operations are performed. When precedence is equal the operations are performed left to right. Parentheses can be used by the programmer to override the order.

Figure A-41 shows the use of a CPT instruction. The mathematical operations are performed when contact I:012/10 is true. The result of the operation is stored in the destination address (N76:20).

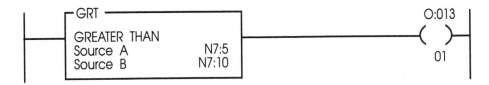

Figure A-41. Use of a compute (CPT) instruction.

Add

The add instruction (ADD) is used to add two values together (source A + source B). The result is put in the destination address. Figure A-42 shows an example of an add instruction. If contact I:012/10 is true, the add instruction will add the number from source A (N7:3) and the value from source B (N7:4). The result will be stored in the destination address (N7:20). The effects on the status bits are shown in Figure A-43.

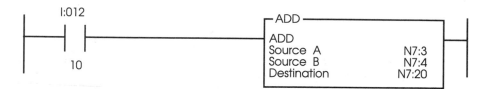

Figure A-42. Use of an add instruction.

Bit	CPU Action:
Carry (C)	1 if carry generated, otherwise resets
Overflow (V)	1 if overflow occurs, otherwise resets
Zero (Z)	1 if result = zero, otherwise resets
Sign (S)	1 if result is negative, otherwise resets

Figure A-43. Effects of an add instruction on the status bits.

Subtraction

The subtraction instruction (SUB) is used to subtract two values. The subtract instruction subtracts source B from source A. The result is stored in the destination address. Figure A-44 shows the use of a subtract instruction (SUB). If contact I:012/10 is true, the SUB instruction is executed. Source B is subtracted from source A; the result is stored in destination address N7:20. The effect upon status bits is shown in Figure A-45.

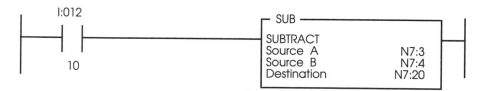

Figure A-44. This figure shows the use of a subtract (SUB) instruction.

Bit	CPU Action:
Carry (C)	1 if borrow occurs, otherwise a 0
Overflow (O)	1 if underflow occurs, otherwise a 0
Zero (Z)	1 if result = zero, otherwise a zero
Sign (S)	1 if result is negative, otherwise a 0

Figure A-45. How status bits are affected by a subtract instruction.

Multiply

The multiply instruction (MUL) is used to multiply two values. The first value, source A, is multiplied by the second value, source B. The result is stored in the destination address. Source A and source B can be either values or addresses of values. Figure A-46 shows the use of a multiply instruction. If contact I:012/10 is true, source A (N7:3) is multiplied by source B (N7:4) and the result is stored in destination address N7:20. Status bits are set according to Figure A-47.

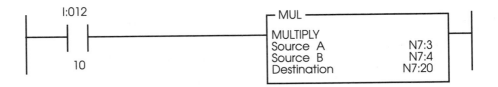

Figure A-46. Use of a multiply (MUL) instruction.

Bit	CPU Action:
Carry (C)	Always resets
Overflow (V)	1 if overflow occurs, otherwise a 0
Zero (Z)	1 if result = zero, otherwise a 0
Sign (S)	1 if result is negative, otherwise a 0

Figure A-47. How a multiply instruction affects status bits.

Divide

The divide instruction (DIV) is used to divide two values. Source A is divided by source B and the result is placed in the destination address. The sources can be values or addresses of values. Figure A-48 shows the use of a divide instruction. If contact I:012/10 is true, the divide instruction will divide the value from source A (N7:3) by the value from source B (N7:4). The result is stored in destination address N7:20. Figure A-49 shows how a divide instruction affects the status bits.

Figure A-48. Use of a divide (DIV) instruction.

Bit	CPU Action:
Carry (C)	Always resets
Overflow (V)	1 if overflow occurs or if division by 0, otherwise a 0
Zero (Z)	1 if result = zero, otherwise a 0, undefined if overflow is set
Sign (S)	1 if result is negative, otherwise a 0, undefined if overflow set

Figure A-49. How the status bits are affected by the divide instruction.

Negate

The negate instruction (NEG) is used to change the sign of a value. If it is used on a positive number, it makes it a negative number. If it is used on a negative number, it will change it to a positive number. Remember that this instruction will execute everytime the rung is true. Use transitional contacts if needed. The use of a negate instruction is shown in Figure A-50.

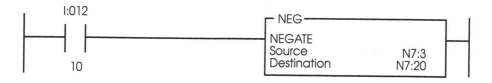

Figure A-50. Use of a negate (NEG) instruction.

Bit	CPU Action:
Carry (C)	Sets if the operation generates a carry
Overflow (V)	Sets if an overflow is generated, otherwise resets
Zero (Z)	Sets if result is zero, otherwise resets
Sign (S)	Sets if result is negative, otherwise resets

Figure A-51. Effect of a negate instruction on status bits.

If contact I:012/10 is true, the value in source A (N7:3) will be given the opposite sign and stored in destination address N7:20. Figure A-51 shows how status bits are affected by a negate instruction.

Clear Instruction

The clear instruction (CLR) is used to set all bits of a word to zero. The destination must be an address. An example is shown in figure A-52. If contact I:012/10 is true, the negate instruction will clear the value in address N7:3. All bits in word N7:3 will be set to zero. The effect on status bits is shown in Figure A-53.

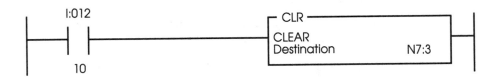

Figure A-52. Use of a clear (CLR) instruction.

Bit	CPU Action:
Carry (C)	Always resets
Overflow (V)	Always resets
Zero (Z)	Always sets
Sign (S)	Always resets

Figure A-53. Effect of a clear instruction on status bits.

Square Root

The square root instruction is used to find the square root of a value. The result is stored in a destination address. The source can be a value or the address of a value.

Figure A-54 shows the use of a square root instruction. If contact I:012/10 is true, the SQR instruction will find the square root of the value of the number found at the source address N7:3. The result will be stored at destination address N7:20. The effect on status bits is shown in Figure A-55.

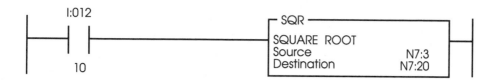

Figure A-54. Use of a square root (SQR) instruction.

Bit	CPU Action:
Carry (C)	Always resets
Overflow (V)	Sets if an overflow generated during floating-point to integer conversion, otherwise resets
Zero (Z)	Sets if result is zero, otherwise resets
Sign (S)	Always resets

Figure A-55. Effect of a SQR instruction on the status bits.

Standard Deviation

The standard deviation instruction (STD) is used to find the standard deviation of a set of values. It stores the result in a destination address. It is used extensively in process control applications. The instruction is executed only on a false to true transition.

Figure A-56 shows an example of the use of a STD instruction. There are several values that the programmer must enter. File is an address that contains the first value of the set to be included in the calculation. The destination is the address where the result of the calculation will be stored. Control is the address of the control structure in the control area (R) of the CPU memory. The CPU uses this information to perform the instruction. Length is the number of values that will be used in the calculation (number of file elements, 0 to 1000). Position points to the element that the instruction is currently using.

In Figure A-56, if contact I:012/10 is true, the STD instruction will execute. When the STD instruction is enabled, the .EN bit for R6:0 is set. This turns on output O:010/05. The instruction uses the values from elements N7:1, N7:2, N7:3, and N7:4 to calculate the standard deviation. The result is stored in destination address N7:0. When the instruction is done the done bit (.DN) is set, turning output O:010/07 on.

The effect of the STD instruction on the status bits is shown in Figure A-57.

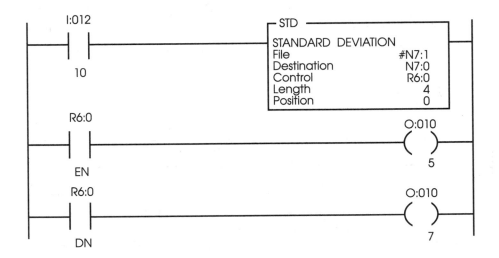

Figure A-56. Use of a standard deviation instruction.

Bit	CPU Action:
Carry (C)	Always resets
Overflow (V)	1 if overflow occurs, otherwise a 0
Zero (Z)	1 if result = zero, otherwise a 0
Sign (S)	Always resets

Figure A-57. Effect of the STD instruction on the status bits.

Average Instruction

The average instruction (AVE) is a file instruction. It is used to find the average of a set of values. The AVE instruction calculates the average using floating-point regardless of the type specified for the file or destination. If an overflow occurs, the CPU aborts the calculation. In that case the destination remains unchanged.

The position points to the element that caused the overflow. When the .ER bit is cleared, the position is reset to zero and the instruction is recalculated. Everytime there is a low-to-high transition the value of the current element is added to the next element. The next low-to-high transition causes the current element value to be added to the next element, and so on. Every time another element is added, the position field and the status word are incremented.

Figure A-58 shows how an average instruction (AVE) is used in a ladder diagram. The programmer must provide certain information while programming. The file is the address of the first element to be added and used in the calculation. The destination is the address where the result will be stored. This address can be floating-point or integer. The control is the address of the control structure in the control area (R) of CPU memory. The CPU uses this information to run the instruction. Length is the number of elements to be included in the calculation (0 to 1000). Position points to the element that the instruction is currently using.

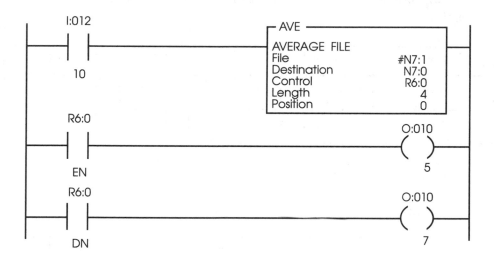

Figure A-58. Use of an average (AVE) instruction.

Logical Operator

There are several logical instructions available. They can be very useful to the innovative programmer. They can be used, for example, to check the status of certain inputs while ignoring others.

And

There are several logical operator instructions available. The AND instruction is used to perform an AND operation using the bits from two source addresses. The bits are ANDed and a result occurs. See Figure A-59 for a chart that shows the results of the four possible combinations. An AND instruction needs two sources (numbers) to work with. These two sources are ANDed and the result is stored in a third address.

Figure A-60 shows an example of an AND instruction. Addresses D9:3 and address D10:4 are ANDed. The result is placed in destination address D12:3. Examine the bits in the source addresses so that you can understand how the AND produced the result in the destination.

Source A	Source B	Result
0	0	0
1	0	0
0	1	0
1	1	1

Figure A-59. Results of an AND operation on the four possible bit combinations.

Source A D9:3	0	0	0	0	0	0	0	0	1	0	1	0	1	0	1	0
Source B D10:4	0	0	0	0	0	0	0	0	1	1	1	0	1	0	1	1
Destination D12:3	0	0	0	0	0	0	0	0	1	0	1	0	1	0	1	0

Figure A-60 Result of an AND on two source addresses. The ANDed result is stored in address D12:3.

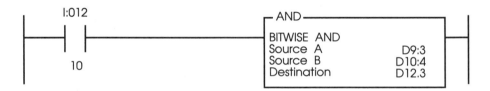

Figure A- 61. Use of an AND instruction. If input I:012/10 is true, the AND instruction executes. The number in address D9:3 is ANDed with the value in address D10:4. The result of the AND is stored in address D12:3.

NOT

NOT instructions are used to invert the status of bits. A one is made a zero and a zero is made a 1. See Figures A-62, A-63, and A-64.

Source	Result
0	1
1	0

Figure A-62. Results of a NOT instruction on bit states.

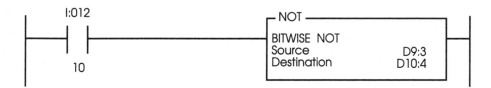

Figure A-63. Use of a NOT instruction. If input I:012/10 is true, the NOT instruction executes. The number in address D9:3 is NOTed with address D10:4. The result is stored in destination address D12:3.

Source D9:3	0	0	0	0	0	0	0	0	1	0	1	0	1	0	1	0
Destination D10:4	0	0	0	0	0	0	0	0	0	1	0	1	0	1	0	1

Figure A-64. What would happen to the number 0000000010101010 if a NOT instruction were executed. The result is shown in destination address D10:4.

OR

Bitwise OR instructions are used to compare two the bits of two numbers. See Figures A-65, A-66, and A-67 for examples of how the instruction functions.

Source A	Source B	Result
0	0	0
1	0	1
0	1	1
1	1	1

Figure A-65. Result of an OR instruction on bit states.

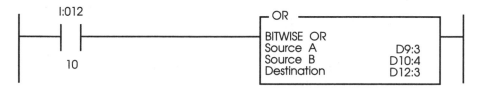

Figure A-66. How an OR instruction can be used in a ladder diagram. If input I:012/10 is true, the OR instruction executes. Source A (D9:3) is ORed with source B (D10:4). The result is stored in the destination address.

Source A D9:3	0	0	0	0	0	0	0	0	1	0	1	0	1	0	1	0
Source B D10:4	0	0	0	0	0	0	0	0	1	1	1	0	1	0	1	1
Destination D12:3	0	0	0	0	0	0	0	0	1	1	1	0	1	0	1	1

Figure A-67. Result of an OR instruction on the numbers in address D9:3 and address D10:4. The result of the OR is shown in the destination address D12:3.

Exclusive Or

Bitwise exclusive or instructions (XOR) are used to compare the bits of two numbers. The result of the XOR is placed in the destination address. Figures A-68, A-69, and A-70 show how the instruction is used.

Source A	Source B	Result
0	0	0
1	0	1
0	1	1
1	1	0

Figure A-68. How the XOR instruction evaluates bit states.

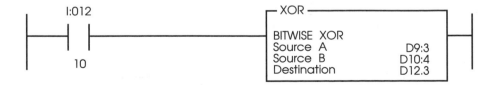

Figure A-69. Use of a XOR instruction. If contact I012/10 is true, source A is exclusive ORed with source B. The result is stored in address D12:3. Source A and source B are not modified.

Source A D9:3	0	0	0	0	0	0	0	0	1	0	1	0	1	0	1	0
Source B D10:4	0	0	0	0	0	0	0	0	1	1	1	0	1	0	1	1
Destination D12:3	0	0	0	0	0	0	0	0	0	1	0	0	0	0	0	1

Figure A-70. How an XOR instruction would evaluate two sources. The result is stored in destination address D12:3. The source addresses are not modified.

Number System Conversion

From BCD

The from BCD instruction (FRD) is used to convert a BCD number to its binary equivalent (see Figure A-71). The source number (a BCD value) is converted and its binary equivalent is stored in the destination address.

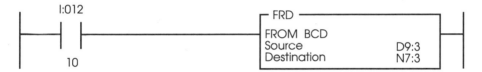

Figure A-71. Use of a FRD instruction. If contact I:012/10 is true, the instruction executes. Source address (D9:3) contains a BCD value. The FRD instruction converts it to its binary equivalent. The result is stored in the destination address (N7:3)

To BCD

The to BCD instruction (TOD) is used to convert a binary number to a BCD equivalent. The source address contains a binary number. The TOD converts it to a BCD value and stores it to the destination address.

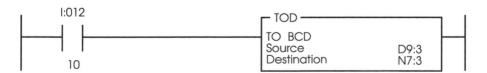

Figure A-72. Use of a TOD instruction.

There are a wealth of other instructions available for the AB-5 family of processors. This appendix is intended to make the technician comfortable with some of the basic instructions. The PLC programming manual provides information on many additional instructions.

Appendix B

Input Device Symbols

Normally Open

Normally Closed

Normally Open - Held Closed

Normally Closed - Held Open

Liquid Level Switches

Closes on Rise

Opens on Rise

Temperature Switches

Closes on Rise

Opens on Rise

Pressure or Vacuum Switches

Closes on Increase

Opens on Increase

Flow Switches

Closes on Increase

Closes on Decrease

Time Delay Contacts

Normally Open
Time Delay Opening

Normally Closed
Time Delay Closing

Normally Open
Time Delay Closing

Normally Closed
Time Delay Opening

Output Device Symbols

Relay

Counter

Timer

Solenoid

Three-Phase Motor

DC Motor

Glossary

A

Ac input module: This is a module that converts a real-world AC input signal to the logic level required by the PLC processor.

Ac output module: Module that converts the processor logic level to an ac output signal to control a real-world device.

Accumulated value: Applies to the use of timers and counters. The accumulated value is the present count or time.

Actuator: Output device normally connected to an output module. An example would be an air valve and cylinder.

Address: Number used to specify a storage location in memory.

Ambient temperature: Temperature that naturally exists in the environment. For example, the ambient temperature of a PLC in a cabinet near a steel furnace is very high.

Analog: Signal with a smooth range of possible values. For example: a temperature that could vary between 60 and 300 degrees would be analog in nature.

ANSI: American National Standards Institute.

ASCII: American Standard Code for Information Interchange. A coding system used to represent letters and characters. Seven-bit ASCII can represent 128 different combinations. Eight-bit ASCII (extended ASCII) can represent 256 different combinations.

Asynchronous communications: Method of communications that uses a series of bits to send data between devices. There is a start bit, data bits (7 or 8), a parity bit (odd, even none, mark, or space), and stop bits (1, 1.5, or 2). One character is transmitted at a time. RS-232 is the most common.

B

Backplane: Bus in the back of a PLC chassis. It is a printed circuit board with sockets that accept various modules.

Baud rate: Speed of serial communications. The number of bits per second transmitted. For example, RS-232 is normally used with a baud rate of 9600. This would be about 9600 bits per second. It takes about 10 bits in serial to send on ASCII character so that a baud rate of 9600 would transmit about 960 characters per second.

Binary: Base two number system. Binary is a system in which 1's and zeros are used to represent numbers.

Binary-coded decimal (BCD): A number system. Each decimal number is represented by four binary bits. For example, the decimal number 967 would be represented by 1001 0110 0111 in BCD.

Bit: Binary digit. The smallest element of binary data. A bit will be either a zero or a one.

Boolean: Logic system that uses operators such as AND, OR, NOR, and NAND. This is the system that is utilized by PLCs, although it is usually made invisible by the programming software for the ease of the programmer.

Bounce: This is an undesirable effect. It is the erratic make and break of electrical contacts.

Branch: Parallel logic path in a ladder diagram.

Byte: Eight bits or two nibbles. (A nibble is 4 bits.)

C

Cascade: Programming technique that is used to extend the range of timers and counters.

CENELEC: European Committee for Electrotechnical Standardization: it develops standards which cover dimensional and operating characteristics of control components.

Central processing unit (CPU): Microprocessor portion of the PLC. It is the portion of the PLC that handles the logic.

Color mark sensor: Sensor that was designed to differentiate between two different colors. They actually differentiate on the basis of contrast between the two colors.

CMOS (complementary metal-oxide semiconductor): Integrated circuits that consume very little power and also have good noise immunity.

Compare instruction: PLC instruction that is used to test numerical values for "equal", "greater than", or "less than" relationships.

Contact: Symbol used in programming PLCs. Used to represent inputs. There are normally open and normally closed contacts. Contacts are also the conductors in electrical devices such as starters.

Contactor: Special-purpose relay that is used to control large electrical current.

CSA (Canadian Standards Organization): Develops standards, tests products and provides certification for a wide variety of products.

Current sinking: Refers to an output device (typically an NPN transistor) that allows current flow from the load through the output to ground.

Current sourcing: Output device (typically a PNP transistor) that allows current flow from the output through the load and then to ground.

D

Dark-on: Refers to a photosensor's output. If the sensor output is on when no object is sensed, it is called a dark-on sensor.

Data highway: This is a communications network that allows devices such as PLCs to communicate. They are normally proprietary, which means that only like devices of the same brand can communicate over the highway. Allen Bradley calls their PLC communication network "Data Highway".

Debugging: Process of finding problems (bugs) in any system.

Diagnostics: Devices normally have software routines that aid in identifying and finding problems in the device. They identify fault conditions in a system.

Digital output: An output that can have two states: on or off. These are also called discrete outputs.

Distributed processing: The concept of distributed processing allows individual discrete devices to control their area and still communicate to the others via a network. The distributed control take the processing load off the "host" system.

Documentation: Documentation is descriptive paperwork that explains a system or program. It describes the system so that the technician can understand, install, troubleshoot, maintain, or change the system.

Downtime: The time a system is not available for production or operation is called down-time. Downtime can be caused by breakdowns in systems.

E

EEPROM: Electrically erasable programmable read only memory.

Energize: Instruction that causes a bit to be a one. This turns an output on.

Examine-off: Contact used in ladder logic. It is a normally closed contact. The contact is true (or closed) if the real-world input associated with it is off.

Examine-on: Contact used in ladder logic programming. Called a normally open contact. This type of contact is true (or closed) if the real-world input associated with it is on.

F

False: Disabled logic state (Off).

Fault: Failure in a system that prevents normal operation of a system.

Flowchart: Used to make program design easier.

Force: Refers to changing the state of actual I/O by changing the bit status in the PLC. In other words, a person can force an output on by changing the bit associated with the real-world output to a 1. Forcing is normally used to troubleshoot a system.

Full duplex: Communication scheme where data flows both directions simultaneously.

G

Ground: Direct connection between equipment (chassis) and earth ground.

H

Half duplex: Communication scheme where data flows both directions but in only one direction at a time.

Hard contacts: Physical switch connections.

Hard copy: Printed copy of computer information.

Hexadecimal: Numbering system that utilizes base 16.

Host computer: Host computer is one to which devices communicate. The host may download or upload programs, or the host might be used to program the device. An example would be a PLC connected to a microcomputer. The host (microcomputer) "controls" the PLC by sending programs, variables, and commands. The PLC controls the actual process but at the direction and to the specifications of the host.

I

IEC (International Electrotechnical Commission): Develops and distributes recommended safety and performance standards.

IEEE: Institute of Electrical and Electronic Engineers.

Image table: Area used to store the status of input and output bits.

Instruction set: Instructions that are available to program the PLC.

Intelligent I/O: PLC modules that have a microprocessor built in. An example would be a module that would control closed-loop positioning.

Interfacing: Connection of a PLC to external devices.

I/O (input/output): Used to speak about the number of inputs and outputs that are needed for a system, or the number of inputs and outputs that a particular programmable logic controller can handle.

IP rating: Rating system established by the IEC that defines the protection offered by electrical enclosures. It is similar to the NEMA rating system.

Isolation: Used to segregate real-world inputs and outputs from the central processing unit. Isolation assures that even if there is a major problem with real-world inputs or outputs (such as a short), the CPU will be protected. This isolation is normally provided by optical isolation.

K

K: Abbreviation for the number 1000. In computer language it is equal to two to the tenth, or 1024.

Keying: Technique to ensure that modules are not put in the wrong slots of a PLC. The user sets up the system with modules in the desired slots. The user then keys the slots to assure that only a module of the correct type can be physically installed.

L

Ladder diagram: Programmable controller language that uses contacts and coils to define a control sequence.

LAN: See Local area network.

Latch: An instruction used in ladder diagram programming to represent an element that retains its state during controlled toggle and power outage.

Leakage current: Small amount of current that flows through load-powered sensors. The small current is necessary for the operation of the sensor. The small amount of current flow is normally not sensed by the PLC input. If the leakage is too great a bleeder resistor must be used to avoid false inputs at the PLC.

LED (light-emitting diode): A solid-state semiconductor that emits red, green, or yellow light or invisible infrared radiation.

Light-on sensor: This refers to a photosensor's output. If the output is on when an object is sensed, the sensor is a light-on sensor.

Linear output: Analog output.

Line-powered sensor: Normally, three-wire senors, although four-wire sensors also exist. The line-powered sensor is powered from the power supply. A separate wire (the third) is used for the output line.

Load: Any device that current flows through and produces a voltage drop.

Load-powered sensor: A load-powered sensor has two wires. A small leakage current flows through the sensor even when the output is off. The current is required to operate the sensor electronics.

Load resistor: A resistor connected in parallel with a high-impedance load to enable the output circuit to output enough current to ensure proper operation.

Local area network (LAN): A system of hardware and software designed to allow a group of intelligent devices to communicate within a fairly close proximity.

LSB: Least significant bit.

M

Machine language: Control program reduced to binary form.

MAP (manufacturing automation protocol): "Standard" developed to make industrial devices communicate more easily. Based on a seven-layer model of communications.

Master control relay (MCR): Hardwired relay that can be deenergized by any hardwired series-connected switch. Used to deenergize all devices. If one emergency switch is hit it must cause the master control relay to drop power to all devices. There is also a master control relay available in most PLCs. The master control relay in the PLC is not sufficient to meet safety requirements.

Memory map: Drawing showing the areas, sizes, and uses of memory in a particular PLC.

Microsecond: A microsecond is one millionth (0.000001) of a second.

Millisecond: A millisecond is one thousandth (.001) of a second.

Mnemonic codes: Symbols designated to represent a specific set of instructions for use in a control program.

MSB: Most significant bit.

N

NEMA (National Electrical Manufacturers Association): Develops standards that define a product, process, or procedure. The standards consider construction, dimensions, tolerances, safety, operating characteristics, electrical rating and so on. They are probably best known for their rating system for electrical cabinets.

Network: System that is connected to for communication purposes.

Node: Point on the network that allows access.

Noise: Unwanted electrical interference in a programmable controller or network. It can be caused by motors, coils, high voltages, welders, and so on. It can disrupt communications and control.

Non-retentive timer: Timer that loses the time if the input enable signal is lost.

Nonvolatile memory: Memory in a controller that does not require power to retain its contents.

NOR: The logic gate that results in zero unless both inputs are zero.

NOT: The logic gate that results in the complement of the input.

O

Octal: Number system based on the number 8, utilizing numbers 0 through 7.

Off-delay timer: This is a type of timer that is on immediately when it receives its input enable. It turns off after it reaches its preset time.

Off-line programming: Programming that is done while not attached to the actual device. For example, a PLC program can be written for a PLC without being attached. The program can then be downloaded to the PLC.

On-delay timer: Timer that does not turn on until its time has reached the preset time value.

One-shot contact: Contact that is only on for one scan when activated.

Operating system: The fundamental software for a system that defines how it will store and transmit information.

Optical isolation: Technique used in I/O module design that provides logic separation from field levels.

OR: Logic gate that results in 1 unless both inputs are 0.

P

Parity: Bit used to help check for data integrity during a data communication.

Peer-to-peer: This is communication that occurs between similar devices. For example, two PLCs communicating would be peer-to-peer. A PLC communicating to a computer would be device-to-host.

PID (Proportional, integral, derivative) control: Control algorithm that is used to closely control processes such as temperature, mixture, position, and velocity. The proportional portion takes care of the magnitude of the error. The integral takes care of small errors over time. The derivative compensates for the rate of error change.

PLC: Programmable logic controller.

Programmable controller: A special-purpose-computer. Programmed in ladder logic. It was also designed so that devices could be easily interfaced with it.

Pulse modulated: Turning a light source on and off at a very high frequency. In sensors the sending unit pulse modulates the light source. The receiver only responds to that frequency. This helps make photo-sensors immune to ambient lighting.

R

Rack: PLC chassis. Modules are installed in the rack to meet the user's need.

Radio frequency (RF): Communications technology in which there is a transmitter/receiver and tags. The transmitter/ receiver can read or write to the tags. There are active and passive tags available. Active tags are battery powered. Passive tags are powered from the RF emitted from the transmitter. Active tags have a much wider range of communication. Either tag can have several "K" of memory.

RAM (random access memory): Normally considered user memory.

Register: Storage area. It is typically used to store bit states or values of items such as timers and counters.

Retentive timer: Timer that retains the present count even if the input enable signal is lost. When the input enable is active again, the timer begins to count again from where it left off.

Retroreflective: Photosensor that sends out a light which is reflected from a reflector back to the receiver (the receiver and emitter are in the same housing). When an object passes through it breaks the beam.

RF (radio frequency): See radio frequency.

ROM (read-only memory): This is operating system memory. ROM is nonvolatile. It is not lost when the power is turned off.

RS-232: Common serial communications standard. This standard specifies the purpose of each of 25 pins. It does not specify connectors or which pins must be used.

RS-422 & RS-423: Standards for two types of serial communication. RS-422 is a balanced serial mode. This means that the transmit and receive lines have their own common instead of sharing one like RS-232 Balanced mode is more noise immune. This allows for higher data transmission rates and longer transmission distances. RS-423 uses the unbalanced mode. Its speeds and transmission distances are much greater than RS-232 but less than RS-422.

RS-449: Electrical standard for RS-422/RS-423. It is a more complete standard than the RS-232 It specifies the connectors to be used also.

Rung: Group of contacts that control one or more outputs. In a ladder diagram it is the horizontal lines on the diagram.

S

Scan time: Amount of time it takes a programmable controller to evaluate a ladder diagram. The PLC continuously scans the ladder diagram. The time it takes to evaluate it once is the scan time. It is typically in the low-millisecond range.

Sensitivity: Refers to a devices ability to discriminate between levels. If its a sensor it would relate to the finest difference it could detect. If it were an analog module for a PLC, it would be the smallest change it could detect.

Sensor: Device used to detect change. Normally they are a digital device. The outputs of sensors change state when they detect the correct change. Sensors can be analog or digital in nature. They can also be purchased with normally closed or normally open outputs.

Sequencer: Instruction type that is used to program a sequential operation.

Serial communication: Sending of data one bit at a time. The data is represented by a coding system such as ASCII.

Speech modules: Used by a PLC to output spoken messages to operators. The sound is typically digitized human speech stored in the module's memory. The PLC requests the message number to play it.

T

Thermocouple: A thermocouple is a sensing transducer. It changes a temperature to a current. The current can then be measured and converted to a binary equivalent that the PLC can understand.

Thumbwheel: Device used by an operator to enter a number between 0 and 9. Thumbwheels are combined to enter larger numbers. Thumbwheels typically output BCD numbers to a device.

Timer: Instruction used to accumulate time until a certain value is achieved. The timer then changes its output state.

TOP (technical and office protocol). Communication standard that was developed by Boeing. Based on the contention access method. The MAP standard is meant for the factory floor and TOP is meant for the office and technical areas.

Transitional contact: Contact that changes state for one scan when activated.

True: This is the enabling logic state. Generally associated with a "one" or "high" state.

U

UL (Underwriters Laboratory): Organization that operates laboratories to investigate systems with respect to safety.

User memory: Memory used to store user information. The user's program, timer/counter values, input/output status, and so on, are all stored in user memory.

V

Volatile memory: Memory that is lost when power is lost.

W

Watchdog timer: Timer that can be used for safety. For example, If there is an event or sequence that must occur within a certain amount of time, a watchdog timer can be set to shut the system down in case the time is exceeded.

Word: Length of data in bits that a microprocessor can handle. For example, a word for a 16-bit computer would be 16 bits long, or two bytes. A 32-bit computer would have a 32-bit word.

Index

A

Access methods   204
Active-high   10
Allen Bradley PLC-2 counters   78
Allen Bradley PLC-2 timers   68
Analog   259
Analog input modules   125
Analog output   24
Analog output modules   127
Analog sensors   96
AND   52
ANSI   259
Applications   27
Area control   199
Arithmetic instructions   142
ASCII   13, 195, 259
ASCII modules   129
Asynchronous communications   259

B

Backplane   259
Bar-code modules   133
Baseband   202
Baud rate   259
Binary   35, 259
Binary coded decimal   37, 260
Bipolar module   125
Bit   260
Bleeder resistor   121
Boolean   260
Bounce   260
Branch   54, 260
Broadband   202
Burden current   107
Byte   36, 260

C

Capacitive sensors   104
Cascade   260
Cascading timers   77
Cell level   193

CENELEC   260
Central processing unit   7, 260
Channels   18
Closed-loop position control   130
CMOS   260
CMOS-RAM   10
Coaxial cable   202
Coil   47
Collision detection   205
Color mark sensor   99, 260
Communication modules   128
Compare instructions   144
Compensation   109
Contact   46, 260
Contact sensor   96
Contactor   260
Counters   77-84
CPU   17, 60
CSA   260
CSMA/CD   205
Current ratings   25
Current sinking   260
Current sourcing 260
Current specifications   122

D

Dark sensing   97
Dark-on   261
Data   190
Data highway   261
Debounce   23
Decimal   34
Delay-off timing   66
Delay-on   66
Device level   192
Diffuse   98
Digital input modules   120
Digital modules   120
Digital output   24, 261
Digital output modules   122
Digital sensors   96, 97
Distributed processing   261

Documentation   15
Down counters   78
Drum controller   164
Dumb terminal   13

E

EEPROM   12, 261
Electrical noise   7, 217
Electronic field sensors   100
Enclosures   212
Energize   261
EPROM   10
Examine-off   261
Examine-on   261

F

Fail-safe   50
FDDI   203
Fiber optic   128, 203
Fiber optic sensors   99
Field sensors   100
Flow diagram   85
Flow diagram symbols   85
Flowchart   261
Force   262
Frequency-division multiplexing   202
Full duplex   262
Fuzzy logic   135, 177

G

Gate sensor   95, 114
Gould Modicon counters   80
Gould Modicon timers   70
Grounding   215, 216, 262
Grounding guidelines   216

H

Half duplex   262
Hand-held programmer   14
Handshaking   23
Hardwired control   3
Hexadecimal   38, 262
High speed counter modules   122
High-density modules   125
Host level   208
Host-link modules   128

Hysteresis   103

I

I/O image table   9
I/O modules   120
IEC   262
IEEE   262
Image table   9, 262
Immediate instructions   57
Inductive sensors   101, 102
Industrial terminals   13
Input modules   23
Input section   17
Installation (PLC)   212
Isolation   20, 23, 263

L

Ladder diagram   2, 47, 263
Ladder logic   4, 46
Laser sensors   100
Latch   263
Latching instructions   58
Leakage current   106, 121, 263
Least significant bit   36
Levels of plant communication   190
Light sensing   97
Line-powered sensors   107, 263
Load   263
Load-powered sensor   105, 263
Local area network   199, 264

M

Manufacturing automation protocol
      (MAP)   205, 264
Manufacturing message specification
      (MMS)   207
MAP   205, 264
Master control relay   59, 264
Memory   8, 264
Memory Map   10, 264
Microcomputer   15

N

NEMA   212, 264
Nibble   36
Noise   7, 217, 218, 264

Noise isolation   218
Noise suppression   217
Non-retentive timer   264
Noncontact sensors   96
Nonretentive   68
Nonretentive timers   68
Nonvolatile memory   10, 264
Normally closed   46, 49, 51
Normally open   46
NPN   108

O

Octal   37, 265
Off-delay timer   67, 98, 265
Off-line programming   15, 265
Omron counters   80
Omron timers   71
On-delay timer   98, 265
On-line   15
One-shot contact   265
Open-loop position control   129
Operating system   265
Operating system memory   9
Optical isolation   20, 23, 265
Optical sensors   97
Output image table   25
Output modules   24

P

Parity   265
Peer-to-peer   129, 265
Photodetector   97
PID   265
PID modules   133
PLC applications   27
PLC installation   212
PNP sensor   108
Position control modules   129
Power Supply   16
Preset value   67
Primitive communication   193
Programmable controller   3, 5
Programming devices   12
Pseudocode   86

R

Rack   17, 19, 266
Radio frequency   136, 203, 266
Radio-frequency modules   136
Random access memory (RAM)   10, 266
Read-only memory (ROM)   9
Reflective sensors   98
Relays   25
Remote I/O modules   127
Resolution   109, 127, 126
Response time   106
Retentive timer   68, 266
Retroreflective sensors   99, 266
ROM   266
RS-232   194, 195, 266
RS-422   195
RS-423   195
RS-449   196
RS-485   196
RS-422   266
RS-423   266
RS-449   266
RTD   111
Rung   266

S

SCADA   197
Scan   47
Scan cycle   48
Scan time   60, 266
Sensing distance   102, 104
Sensor   23, 46, 267
Sensor applications   94, 113
Sensor installation   112
Sensor wiring   105
Sensors   94-118
Sequencer instructions   166
Sequential control   164
Serial communication   267
Shift resister programming   169
Siemens Industrial Automation counters   83
Siemens Industrial Automation timers   74
Sinking   108, 260

Slots 18
Snubbing 218
Sourcing 108
Speech modules 138, 267
Square D counters 82
Square D timers 73
Stage programming 170
Start/stop circuit 56
State logic 184
Step programming 174
Strain gage 111
System wiring 213

T

Thermistor 111
Thermocouple 108, 110, 267
Thru-Beam 98, 99
Thumbwheel 37, 267
Time base 67
Time-division multiplexing 202
Timer 66-77, 267
Token passing 204
Topology 200
Transitional contact 267

Transitional contacts 57
Troubleshooting 4, 14, 15, 219
Twisted pair 128, 202

U

Ultrasonic sensors 100
Unipolar modules 125
Up counters 78
Up/down counter 78
Upload/download 15
User memory 9, 267

V

Vision modules 131
Volatile memory 268

W

Watchdog timer 268
Wiring guidelines 214
Word 36, 268

Z

Zone control logic instructions 59